SOCIAL BY NATURE

SOCIAL BY NATURE

The Promise and Peril of Sociogenomics

CATHERINE BLISS

STANFORD UNIVERSITY PRESS
STANFORD, CALIFORNIA

Stanford University Press
Stanford, California

Printed in the United States of America on acid-free, archival-quality paper

Library of Congress Cataloging-in-Publication Data

Names: Bliss, Catherine, author.
Title: Social by nature : the promise and peril of sociogenomics / Catherine Bliss.
Description: Stanford, California : Stanford University Press, 2018. |
   Includes bibliographical references and index.
Identifiers: LCCN 2017007864 (print) | LCCN 2017019776 (ebook) | ISBN
   9781503603967 (e-book) | ISBN 9780804798341 | ISBN 9780804798341 (cloth :
   alk. paper)
Subjects: LCSH: Genomics—Social aspects. | Human behavior genetics. | Human
   genetics—Social aspects.
Classification: LCC QH457.5 (ebook) | LCC QH457.5 .B54 2018 (print) | DDC 572.8/6—dc23
LC record available at https://lccn.loc.gov/2017007864

Typeset by Classic Typography in 10/14 Minion Pro

For Dad

CONTENTS

SOCIAL BY NATURE

INTRODUCTION

"Gangsta Gene" Identified in US Teens[1]

Single Forever? It's in Your Genes[2]

Happiness Is in Your DNA; and Different Races May Have Different Propensities for It[3]

Sexual Orientation Not a Choice, but Is Influenced by Genetics[4]

Everyday more and more studies emerge that tell us about the genetics of who we are. From benign qualities like optimism and intelligence to darker attributes like the tendency toward rage, sexual promiscuity, and abuse, genetics is the new trope for knowing what makes us tick. And this is no longer simply a matter of medical issues or physiological traits like hypertension, depression, and bone density. Our genetics are increasingly said to be the cause of our fundamental beliefs and behaviors; if humans consistently do it, it must be in the genes.

As familiar as it may sound, the scientific drivers of this phenomenon are new. Despite centuries of nature-nurture debates and the long history of applying genetics to behavior, technology sophisticated enough to assess genetic causes for complex human traits and behaviors only emerged a little over a decade ago. Since the mapping of the human genome in the year 2000, scientists have gone from studying single genes and single-gene disorders to multifactorial "polygenic" traits and common disease. More recently, a cadre of scientists hailing from a range of scientific fields have trained their attention on nondisease matters. They have decided to tackle what they see as the Big Questions of human action: choice, desires, and thought.

The avalanche of research has been expeditious. In under a decade, scientists have teamed up across vast disciplinary divides to find innovative ways to combine traditional population genetic and demographic approaches with cutting-edge genomic methods, such that any human trait that demonstrates heritability can be analyzed genomically. They have formed informal research networks and formal scientific consortia that churn through the universe of known behaviors, subjecting trait after trait to the test. Researchers have leveraged shared strengths and built novel partnerships in which disciplinary approaches can be fused and synergistically reconfigured. The result has been an all-out transformation in how we do science, how we investigate what it means to be human, and how we understand the meaning of life.

A New Science Emerges

A new science has emerged as the leader in research and development along these lines: social genomics. Social genomics pairs the twin gold standards of genomic methodology (genomewide association study, or GWAS, and structural DNA analysis) with basic social research in order to find genetic causes for nondisease social phenomena such as educational attainment, gang membership, life satisfaction, and debt. The field is remarkably led by economists, political scientists, and sociologists who are new to genetics. These newcomers are directing studies in efforts to revolutionize genomic research methods so that social outcomes can be linked to genetic traits. The science has rapidly gained ground in public health and the public sphere, popularized as "genoeconomics," "genopolitics," and "genosociology." And it has been successfully institutionalized in the form of global consortia and doctoral traineeships, headlining in venues such as *Science*, the *American Journal of Public Health*, NPR, and the *New York Times*, and garnering support from educators and public officials enthusiastic to apply social genomics findings and technologies.[5] With its rapidly expanding support base of institutes such as the National Institute of Justice and the National Institute of Child Health and Human Development, and its leadership in the genomics of nondisease phenomena like personality and intelligence, social genomics has made its mark as the pioneer in the genomics of behavior.

At the same time, social genomics has risen to power largely under the radar of mainstream genomics. Though the science has brought genomics to bear on an array of traits that are usually stigmatized in members of vulnerable populations—"social phenotypes" like aggression and impulsivity[6]—it has yet to provoke the kind of Establishment critique that prior attempts to characterize behavior have. Unlike the IQ debates of the 1970s, the fervent science-spanning response to *The Bell Curve* of the mid-1990s, or the 2005 biomedical TKO of geneticist Bruce Lahn's racial brain-size claims, there have been no letters to the editor, no podcast rants, and nary an angry phone call. Social genomics is navigating the frontier seas of sociopolitical controversy with little need for a life vest—begging the question of how this new science can take hold in such a way.

Sociological field analysis provides a provocative lens for thinking about these issues. Analysts have long recognized that emergent scientific fields possess unique properties. While all sciences are defined by their specific expertise vis-à-vis other disciplines and professions, and their location in a historically situated institutional ecology,[7] new fields are in a constant state of jurisdiction management, where members compete with other members and with those external to the field to protect, enhance, transform, and enlarge the field's authority.[8] Creating a new field involves controlling professional membership and expertise but also demarcating an independent identity from parent fields and others outside the field.[9] Researchers call this process autonomization.[10]

Studies have shown that autonomization has a special relationship to controversy.[11] As autonomizing fields work to distinguish their expertise from those outside their community, scientists often pursue controversial lines of research in order to set themselves apart from other members of their field or the arena of science that it occupies so that they can compete for a place in the limelight.[12] New problematics and approaches can afford researchers access to newly available resources, social and symbolic capital, expertise over new jurisdictions, and even ethical legitimacy.[13] This can propel fields into the public eye, garnering them notoriety for intellectual bravery. A sort of "new field cheer" can overtake a science, giving it the momentum to slide by the usual critical firing line.

Emergent science can also take the form of intellectual movements, or collective attempts to pursue research in the face of resistance from others in the science community.[14] These controversial lines of research can be seen as helping scientists shine a critical light on the ethical crises that scientists themselves are facing.[15] So even when sciences do not fully autonomize or institutionalize, their public activity can allow scientists to act upon a landscape of moral possibilities to shape mainstream political debates and ideologies.[16]

In a world where sociality is increasingly formed around genomic data, all scientists share a responsibility to enlighten the public on the ties between genes and the environment.[17] Thinking about the implications of genomic data is no longer just a job for genomicists.[18] Nevertheless, the social sciences from which much of social genomics arises are established fields that frequently maintain their jurisdiction by creating alternatives to, if not downright discounting, genetic explanations.[19] Economists have asked what social circumstances encourage us to make the decisions we do in terms of markets and resources, work and education, and a whole host of other issues. Political scientists have wondered why we aggregate as we do and how different forms of government affect us. Sociologists have tried to understand how we are socialized into the people we are, how we form groups and interact in them. This is by no means an exhaustive list of topics that social scientists study, but it presents the overall causal framework that they work from, a system of cause and effect that moves from social factors to individual and collective behavior. This book thus asks, Why is social genomics emerging here and now? What is "in it" for the scientists leading the way? Is it all about publicity and controlling debates? In sociological terms, what are the political, institutional, and scientific motivations that are currently hidden within the supposed legacy of relations between the natural and social sciences?

Another issue at play is today's global charge to create multidisciplinary "team science." Despite the fact that disciplines and programs continue to hire their own,[20] research has shown that academic institutions and research agencies, such as the European Union (EU), National Institutes of Health (NIH), National Science Foundation (NSF), and American Association of Colleges and Universities (AACU), are pouring billions into multidisciplinary projects and sites in order to promote interdisciplinarity and transdisciplinarity in the sciences,[21] motivating some to argue that there is a "limited usefulness of terms such as 'public' and 'private' or 'global,' 'regional,' and 'national' to describe the actual structure of most collaborations, their emergence and outcomes."[22] These models promote a move away from field-based expertise in the interest of tackling specific critical social problems, such as global warming and obesity.

Despite their rejection of the pursuit of science for science's sake, it is still unclear whether team science models are truly inter- or transdisciplinary in practice, versus whether they promote certain disciplinary interests over others.[23] Much research has emphasized the hierarchical structure of the sciences, and the domination of "hard science" concepts and methods in multidisciplinary contexts.[24] But sciences of similar stature have also been shown to successfully "bidirectionally" converge around shared methods and problems, or political goals.[25] Thus this book also asks, What does the multidisciplinary terrain of social genomics look like? How do members of this science deliberate between basic theories and procedures, and how do they seek and find institutional support? Also, to what extent is social genomics fostering lasting ideological and institutional structures? Is social genomics just one instance of a fleeting trend or is it indicative of a new form of science?

Answers to these questions will expose how the science is taking hold in the broader science community, who is rewarded in this arena, and who stands to benefit from it.

The Genetic Politics of Difference

This book is also about the ways in which emerging genetic sciences redefine societal notions of human difference—including but not limited to America's major axes of difference: race, gender, and sexuality.[26] The ethical, legal, and social implications of human genetic research, including how experts use genetic research and how their uses impact racial minorities, women, and LGBTQI (lesbian, gay, bisexual, transgender, queer or questioning, and intersexual) individuals, have become a leading priority area in public health and governance,[27] and have produced a substantial body of research on the science's detrimental influence over the general population.[28] Historical research has shown that training a genetic lens on human traits and behaviors has traditionally led to deterministic notions of difference and to unethical and unscientific research applications, such as when early twentieth-century European and American governmental and political organizations used genetic findings to produce systematic eugenics programs, anti-immigration and forced sterilization campaigns, and the Nazi "Racial Hygiene" program.[29] In the wake of World War II, as the atrocities of Nazism and Nazi science came to light, geneticists mobilized against prior approaches and applications.[30] But despite advocacy against deterministic models, studies show that the tendency for experts to use genetics to explain variation in human traits and behaviors has only amplified with the availability of genomic technologies.[31] Popular controversies over IQ and gay genes have revealed how genetic models continue to dominate discourse on nondisease human traits and behaviors for which few genomic studies exist.[32] And though leaders of the genomics community have, in certain cases, spoken out against unreplicated claims,[33] social genomics has become the new frontier in applied genomics, where judges and preschool administrators have begun implementing it to test criminals, prospective students, and even criminal corpses for prosocial and antisocial traits.[34]

Research has also shown that the application of genomic models to human traits and behaviors has fueled geneticization: the tendency for people to use genetic explanations to describe differences between individual and group traits and behaviors.[35] Since the advent of genomics—the study of DNA sequences— the popular press has ramped up coverage of genetic findings while silencing scientific failures.[36] Meanwhile, deterministic statements have increased since the 2000 publication of the draft map of the human genome.[37] This proliferation of geneticized discourse, matched with the widespread availability of genetic tests, has fostered mass acceptance of genetic essentialism.[38] For example, in the medical arena, where genetic research is depicted without attention to gene-environment interactions,[39] patients increasingly interact with medical practitioners with a sense that the essence of who they are is determined by their personal genetic biology.[40] Similarly, hierarchies in public funding that more readily sponsor DNA research have paved the way for research and healthcare programs that privilege genetic models over environmental explanations.[41]

Studies have proved that both geneticization and the advancement of geneticized models, also known as molecularization,[42] have worse consequences for historically stigmatized groups, especially racial minorities.[43] Historical analyses of the characterization of violence and social resistance in America have shown that genetically deterministic notions have incited stereotypes about the innate inferiority of racial minorities.[44] Studies on contemporary lay beliefs have confirmed that genetically essentialist information increases prejudice and in-group bias, especially for those holding chronic deterministic beliefs.[45] White Americans—who are significantly more likely than black Americans to attribute traits such as athleticism, math performance, drive to succeed, tendency toward violence, intelligence, and sexual orientation to genetics,[46] and to accept genetics as the base cause for mental illness[47]—also are more likely to subscribe to genetic theories that uphold a greater prejudice toward blacks.[48] While exposure to single messages and news headlines about genetics does not always increase people's beliefs in genetic determinism, single messages that link genetics, race, and health do increase racist beliefs.[49] Furthermore, exposure to multiple messages about genetics sparks genetically based racism and racist feelings in those who already hold racist sentiments.[50] Although laypeople tend to understand race in terms of physical characteristics derived from an ambiguous combination of genetic and sociocultural factors, lay audiences that are already showing an increase in racial essentialism are likely to generalize genetic knowledge about physical traits to behavioral ones.[51] This is especially harmful in the case of minority groups that are already stigmatized as having particular diseases.[52]

So far the only social groups seen to potentially benefit from the use of geneticized explanations and models are gay and lesbian adults. One study showed that saying genes were responsible for sexuality made study respondents less blameful of homosexuals and more likely to support gay and lesbian rights.[53] Yet while studies have shown that determinism can motivate support for gay and lesbian domestic partnership and marriage, they have also shown that the same deterministic beliefs overall produce a deeper investment in genetic essentialism, which eventually leads to an increase in societal prejudice.[54]

These findings forecast serious problems for female, minority, and LGBTQI youth. Psychologists have shown that genetic stereotypes can create "stereotype threats," wherein children and adolescents alter their behavior to meet stereotype schema.[55] New genomic characterizations of behavior may therefore create powerful genetically deterministic stereotypes for already stereotyped youth. This can lead to problems with identity construction and social participation throughout childhood and adolescence,[56] setting off a negative chain reaction in self-image, family relationships, and life planning, and eventually leading to worse life outcomes and possibly even shorter life spans.[57]

No one has examined these specific ramifications for disadvantaged youth, nor has anyone examined the impact of tests on youth in nonmedical settings. So far studies on youth have only focused on how genetic test reports make kids feel—whether they increase anxiety and decrease well-being in children,[58] or whether the timing of their disclosure matters to them.[59] None have explored how tests and other forms of genomic knowledge affect youth in the institutional contexts that are programmed to serve them, like schools and juvenile justice centers.[60]

This book seeks to expand these horizons in order to understand what genetics means for human difference inside and outside of biomedicine, and for these especially vulnerable social groups. Are geneticization and molecularization indeed advancing in the public sphere, and are they potentially creating new forms of inequality?

Surprisingly few studies have attempted to explain why genetic models so easily gain mileage in science.[61] Those that have attempted to theorize causes have said that adopting genetic models helps sciences seem more scientific in a publicly appealing way.[62] For example, by molecularizing environmental health science, the environmental sciences have gained scientific and public prestige so that they can better support their political goals of environmental justice while connecting to powerful institutions like the NIH.[63] By molecularizing epidemiology and cognitive science, these fields have increased their public legitimacy while allowing them to expand to new research populations and communities.[64]

Yet this book will shine a light on genetic advocates of a different kind: members of established social science fields that produce findings for institutions that have no present affiliation with genetic science. It will tell us the unique reasons why genetics so widely appeals to nonmedical institutions and experts, and even non-experts.

Discoveries presented in this book—such as the finding that around the world science is attributing innate aggression and antisocial behavior to minority populations, which are then disproportionately investigated, arrested, convicted, and incarcerated, and which face institutionalized discrimination from a range of social institutions that are likely to adopt social genomics; or that men and women are being characterized as fundamentally different down to their DNA—will also contribute a new way of looking at the advancement of genetics. In an era in which medical institutions are partnering with prisons to form genomic databases that can be mined for behavioral traits,[65] and in which racial and sexual profiling are being hotly debated for their utility,[66] the extent to which behavioral traits are geneticized has direct implications for how experts in the broader public will make decisions about individuals over the life course, such as educational placement, college admissions, hiring, and criminal sentencing.[67] This window into uses in the wider public will provide a new theory of geneticization that can illuminate novel causes and consequences pertaining to the twenty-first-century political and historical context.

From Science to Sidewalk

This book is based on a five-year study that began in the office of a nationally renowned public sociologist and ended on the politically charged streets of San Francisco. In 2012, just as my first book, *Race Decoded*,[68] was hitting bookshelves across the nation, I made contact with a leading genosociologist, a professor and dean of a prominent private American university. He immediately introduced me to some of the seminal members of the field—researchers with international projects running in diverse metropolises who were founding professional associations and research institutes dedicated to social genomics. Within weeks, I was planning a trip to Europe to further get to know the global network of scholars with which I was increasingly in frequent communication. I had mapped out where scientists were headquartered, where they typically held research meetings, and where they got their data. I drew up my itinerary, packed my bags, and headed out.

From the earliest moments of my first social genomics meeting, I was struck by the complexity and richness of the field. I saw that the social researchers at the forefront of social genomics were not working alone. Nearly all were novices in genomics, and thus needed help from epidemiologists and medical geneticists who were well versed in genomic theory and method. Conferences were populated evenly by what I would call "genomics collaborators" and card-carrying social scientists, and the balance of hands it took to bring a study to completion seemed weighted in favor of these collaborators.

At my first meeting, I was also made aware of the many institutional representatives that had an interest in the field. There were leaders of national cohort studies who supplied, or were eager to supply, social genomicists with DNA samples. There were directors of funding agencies who sponsored particular meetings and studies, or were there popularizing new midcareer training grants to get social scientists trained in genetics and genomics. There were journal editors soliciting special issues for publication, and university administrators looking to recruit faculty. I came to call these attendees "institutional sponsors," as they were always in some way championing the science and bringing it new opportunities for growth.

Finally, I saw a number of experts who attended simply because they were interested in the potential applications of the research. Some of these people were tied to the cohort studies that provided the samples to social genomicists. Others were academics from far afield who were connected to education, criminal justice, and healthcare. By the sheer numbers, then, you could say that genomic enthusiasts had just as big an interest in the development of this science as any one of its social science originators. I realized that any real understanding of this community would require deep analysis of all of these components.

As I studied on, I found that the diversity of this field was not limited to its population characteristics. Rather, its social contexts, or sites of production, were equally variegated. At the start of my project, I had envisioned going into the field and talking to the requisite experts in their usual venues, just as I had with my prior ethnographic study. In a typical "science study," or qualitative analysis of a field, I would conduct observation and interviews in the academic labs, research centers, and other production sites where scientists bring together the necessary collaborators to meet and plan the collection, processing, and interpretation of genomic data, and to train other scientists, postdoctoral fellows, research associates, and doctoral students. I would also observe the conferences, panels, policy forums, and science advisory meetings that constitute scientists' work. This further observation would focus on venues that fund and facilitate genomics—public health agencies such as the NIH— and those academic, governmental/regulatory, and private sector sites in which scientists present and debate their research.

In contrast, a science study in this present community of research meant meeting scientists in hotel rooms as opposed to labs, watching the planning and execution of genomewide association studies and candidate gene studies via videoconference on Skype, and witnessing the various kinds of scientific decision making, like that which occurs around recruitment and sample coordination, also take place "in silico," via the Web. Though much of any genomic study is "dry lab" work of computer-based data input, sharing, and analysis, social genomics research teams required a wider array of specialized scientists and computing power from all over the world to fill in the blanks of their expertise. More than lab space, social genomicists needed a good Internet connection. Still, scientists would meet for weeklong getaways face-to-face at least once during their collaboration on a study, thus taking me to secluded resorts, ranch homes and condos, and any other quiet forum where leadership teams could work privately in peace.

Similarly, studying social genomics sponsors and adopters meant not just frequenting public health agencies and biotechnology firms, but also going to schools, juvenile justice centers, and childcare facilities where early users like educational administrators, judges, and child welfare officials were applying or thinking about applying the research. And when I set out to observe public engagement, I found my job to be less about monitoring the science column of the *New York Times* and more about tracking pop media's primetime like *This American Life* and *The Colbert Report*.

These new diversified populations and contexts brought a diversification of data. As in my prior work, I planned to rely primarily on observation and interview data that would come from a mix of shadowing participants, asking questions about procedures in progress, informal interviews, observation in which I participated in the interactions being observed, and observation in which I was more of a distant observer. And I knew I would rely on documentary data, including published studies, television and radio transcripts, blogs and websites, grant applications, recruitment and screening instruments, data collection instruments, data analysis and other study protocols, study records, unpublished reports, and published papers or abstracts.[69]

However, studying social genomics required that I incorporate observation data gathered from a whole new set of practices, such as judicial workarounds, teacher training and consultation, and digital platform integration across organizational networks. Likewise, understanding the ways experts access, appraise, and appropriate research evidence required that I gather documentary representations of their institutional policies, protocols, and principles, their lists of populations served and their fiscal sponsorship, as well as related patents, test licenses, and software programs. And with this wider public in mind, popular media and the broader nonmedical web that influenced them were more salient than ever to my research.

I would eventually find that social genomics' vast network of scientific and political entities was part of an even bigger landscape of entities interested in understanding genes as a central rationale for behavior. What was happening in the sciences was merely a microcosm of the larger societal changes afoot. While each of these arenas has had different issues at heart, all have shared a common interest in what I have termed "the sociogenomic paradigm."

The Evolution of Sociogenomics

As argued in *Race Decoded*, since the dawn of the genome era society has been moving toward a sociogenomic paradigm—a view of the world that sees all aspects of human biology and being as an interdependent mix of genetic and social factors.[70] Race is a case in point. The sociogenomic paradigm follows certain philosophical arguments for the genetic reality of race and its continuing use in biomedicine, such as the notion that a cladistic, or tree-based, form of race is compatible with a social constructionist view of it. The basic idea is that even traits for which we have faulty knowledge, and which we only experience in partial ways, can be biologically real.[71] In other words, all genetic and social classifications are arbitrary yet representative of a lived reality that draws from both fonts.

Looking at genomic characterizations of race, my research showed that genome mappers redefined race and health disparities as a product of feedback loops created by genomic ancestry and social ideas about the body, human difference, and health. Scientists responsible for seminal genome projects, who faced pressure from the US public health establishment and an array of experts on race, began to prioritize race-targeted research, minority recruitment into studies, and analysis of genomic health disparities. As a result, large-scale sequencing projects, pharmaceuticals, and genomic research have become ever more racialized, while race has taken on an irrevocably genomic imprimatur.

My research has also shown that genomics has likewise adopted a sociogenomic approach to health disparities, one which defines inequality as of a hybrid "gene-environment" nature requiring immediate expert attention. Yet this approach assumes that any moral indebtedness to racial minorities can be rectified with genomic research inclusion as opposed to economic and political change.[72]

Taken together, these analyses reveal that the sociogenomic paradigm finds a nurturing ground in many domains of society precisely because it has helped with significant ethical problems that scientists, policymakers, and other experts have faced. First, the sociogenomic paradigm has helped scientists think through insurmountable problems of diversity and group labeling, and persistent inequalities in health in the genomic era. Scientists and the public are continually faced with widening gaps in health and life outcomes between self-identified members of different races, with which they are morally charged to find solutions. Ignoring race is feared by many to potentially invite charges of "colorblindism" from minority advocates as well as experts on race.

Yet the sociogenomic paradigm also thrives because it has helped researchers, experts, and advocates interested in becoming leaders in public health to ally with health governance. Health governance institutions have promoted the use of racial classifications to engender inclusion in biomedicine, despite the wealth of studies that have shown that genetic ancestry does not correspond well with lay categories of race. These institutions have not encouraged scientists to devise alternative categories, but rather have increasingly required finer-toothed mechanisms for standardizing the use of social categories of race. Even those scientists with the largest research budgets and intellectual autonomy have adopted the tack of promoting minority inclusion along governmental categorical lines in their efforts to meet public goals for health equity and bring genomics into the fold of public health.

This kind of ethical empowerment is critical in the postgenomic age, "the period after the completion of the sequencing of the human genome [defined by] the advent of whole-genome technologies as a shared platform for biological research across many fields and social arenas," when health equity is the watchword of the day for all scientists.[73] Scientists have fashioned themselves as public reformers and stewards of justice. They maintain that it is their responsibility to be public intellectuals, to be visible in the political mainstream. Scientists hope that the rest of the biosciences, and the broader society, will catch up to them in their understanding of health equity, a view justified by expertise on the most essential piece of the puzzle: gene-environment relations.[74]

In this moment, public outreach and education is a top priority for science writ large. In fact, scientists write their commitments into their grant proposals and seek evidence of social goals in their reviews of others.[75] Scientists engage community groups and community-based representatives on ethical matters in the design phase of study planning.[76] Funding agencies sponsor science advisory panels to broker research proposals between scientists and the public.[77]

And as suggested earlier, interdisciplinary collaboration is a leading priority for all the sciences in this moment. Thus, collaborations in the form of seminars, conferences, and joint publications allow scientists to bridge the chasm between social science and natural science, while providing opportunities for their expertise to shine beyond disciplinary borders. Scientists have become well versed in talking about their own biases and assumptions, and the potential for their identities to affect their science. Owning up to one's position and partiality, prior to launching studies, is a mainstay of the postgenomic posture.

Likewise, local empowerment through knowledge production is a key metric of scientific success. Scientific collectives in the United Kingdom, United States, Canada, and Australia push models such as collective innovation, genomic sovereignty, and multinational partnership in the interest of offering training and supportive outsourcing to the developing world.[78] Individual researchers trade on their knowledge of specific communities to shuttle resources to underserved and vulnerable populations.[79]

Increasingly, then, being a "good scientist" requires the kind of ethical flexibility afforded by the sociogenomic paradigm. As historians of science Sarah Richardson and Hallam Stevens argue:

> For many, postgenomics signals a break from the gene-centrism and genetic reductionism of the genomic age. As scientists narrate the history of the genome sequencing projects, they trace a path from a simplistic, deterministic, and atomistic understanding of the relationship between genes and human characters toward, in the postgenomic era, an emphasis on complexity, indeterminacy, and gene-environment interactions.[80]

The sociogenomic paradigm enables scientists to incorporate new areas of life to study, while potentially moving away from the tarnish of genetic essentialism.

A sociogenomic outlook also melds with the charisma-driven style of thought by which historian and sociologist of science Steven Shapin has characterized the biotechnology industry in the latter half of the twentieth century,[81] and which I described in the case of genomic pioneers in the early twenty-first century, in which principal investigators and project leaders overtly rely on personality and display personal enthusiasm for social agendas to advance their scientific interests. It allows researchers to openly move through biomedicine, science, and society, appealing to a wide array of experts, so that they may align with various nongenomic scientists in the effort to bring a social equality framework into the heart of research.

In the pages that follow, I move from the particularities of bioscience outward to the general framework of popular understanding to further explore these issues of outlook, posture, and worldview. I explore the making of differences and disciplines in their own right, but also in relationship to one another, as essential pillars of our sociogenomic world.

CHAPTER 1

GENES AND THEIR ENVIRONMENTS

A biological hurricane is approaching the social sciences and it's inescapable.

—Sociologist, American University

I begin our journey by providing the historical intellectual context for the emergence of sociogenomic science. Here I move chronologically from early modern biology to genetics, showing that the debate over nature and nurture, or genes and their environments, is a long-standing one with tenacious roots in the foundational assumptions of modern science. I then turn to the resurgence of nature versus nurture precepts in fields like sociobiology, evolutionary psychology, neuropolitics, and neuroeconomics, sciences that have arisen in recent years to great fanfare. I end by discussing their relationship to the present-day gene-environment imperative and sociogenomics, drawing out the paradigm's unique implications for the meaning of politics today.

Genetics and the Origins of Nature versus Nurture

It will likely come as no surprise that the nature-nurture debate that seems to kick up with the launch of each new human science has been with us for centuries now. According to historian and philosopher of science Evelyn Fox Keller, the reason we can't seem to shake this debate is that it consists of a number of different questions all mixed up together, exacerbated by the ambiguities of the language of biology, genetics, and heritability.[1] It's not just that people don't define what they mean when they deploy the terms "nature" and "nurture," but that they advance implicit ideas about the importance of genes and the environment with every use of the terms. And they do so in ways that pit nature against nurture in a hierarchy of causes.

In *The Mirage of a Space between Nature and Nurture*, Keller provides a telling genealogy of the phrase "nature and nurture."[2] She shows that though there were earlier mentions of the words together, like Shakespeare's declarations in *The Tempest* and John Locke's observations when writing about man as a blank slate, before Darwinian evolution the terms "nature" and "nurture" were neither disjointed nor opposed.[3] It wasn't until Francis Galton began interpreting his cousin Charles Darwin's theory of evolution did "nature and nurture" become "nature versus nurture."[4]

Though Darwin introduced the idea of a hard kind of heredity—some internal substance that was passed on from parent to offspring—it was Galton who attributed to genes a "particulate" and invariant character. Darwin and others had drawn a temporal line in the sand between nature and nurture, claiming that nurture kicked in at birth. Galton echoed this idea but to it added that there were innate and acquired elements of a person, the innate being the stronger of the two.[5]

In fact, Galton forever bonded nature to heredity, reversing the trend in science and philosophy up to that point in which heredity was connoted with culture and mores as much as biology. Darwin and Galton thereafter fed off of each other, revising their notions toward a resolutely particulate notion of genetic inheritance. When Galton introduced the notion of inborn mental aptitudes, Darwin was convinced.

We know that Galton became increasingly influential in the late nineteenth century, especially when he issued his science of eugenics.[6] In *Hereditary Genius* (1869),[7] Galton advocated a system that would allow "the more suitable races or strains of blood a better chance of prevailing speedily over the less."[8] Galton also propagated his particulate notion of genetics by ushering in the twin study. Galton used twin studies to examine the families of his contemporaries—those of the British elite—to then argue that mental powers run in families. Galton was soon followed by Edward Thorndike in 1905, who compared fifty pairs of twins also in the efforts to develop a eugenic science.[9] Within a decade, German geneticists Wilhelm Weinberg and Hermann Werner Siemens had formulated the modern twin study approach of comparing monozygotic and dizygotic, or identical and fraternal twins. This approach compares traits in identical and fraternal twins raised in the same family environments to control for any unshared genetic and environmental components. Doing twin studies led researchers to formalize the definition of heritability as the "percentage of variance in a population due to genes,"[10] making for a strict split in the terms "nature" and "nurture."

The devastating history of eugenics with its sterilization campaigns, prona-tal policies, and the Nazi *Lebensproject* and *Rassenhygiene* is widely known. Yet what has often remained shrouded in this history was eugenics' ability—even as eugenics itself transformed—to popularize the mantra that genes and envi-ronment were separate forces, the former being the dominant force. Though not all of Galton's supporters ran with his particulate notion of genetics, one influential scientist did. Karl Pearson, the father of modern statistics, took Gal-ton's terms "nature and nurture" and replaced them with "heredity versus en-vironment." In other words, Pearson took a direction where language slippages would forever be branded into the phrase.[11]

The perpetuation of "nature versus nurture" was also compounded by the development and standardization of IQ tests. In 1908, Sorbonne psychologist Alfred Binet and medical student Theodore Simon finalized an intelligence test to help educators determine which students had learning disabilities.[12] That year, psychologist and prominent eugenicist H. H. Goddard translated the test into English, and soon after Stanford psychologist Lewis Terman standardized the test to "[curtail] the reproduction of feeble-mindedness and in the elimina-tion of an enormous amount of crime, pauperism, and industrial inefficiency."[13] Other influential eugenicists, like Harvard primatologist Robert Yerkes and former British Psychological Association president Cyril Burt, extended IQ tests to new populations, including using them to establish the heritability of IQ in twins and to make interspecies comparisons between humans and apes. The end result was the co-optation of intelligence testing and study by eugenic science.[14]

As genetic science developed into the twentieth century, the nature-nurture schematic was confounded with a lot of popular and scientific misunderstand-ings around the word "genetic." While classical genetics, the science of peas and fruit flies, was only able to measure the difference that having one genetic marker versus another made over generations, the newer science ambitiously sought to characterize how genes acted on the development of traits. In the process, geneticists perpetuated the unproven assumption that phenotypic differences indicated the presence of genetic mutations lurking behind them, wholly responsible for them. Thus the notion that genetic mutations had a 1-to-1 relationship with traits—that genes caused traits—was born.

Though exposure of Nazi genetics and the horrors of its Jewish extermi-nation campaign brought about a reduction in the "positive" and "negative" eugenic strategies of encouraging and discouraging reproduction in particular populations,[15] after World War II the set of practices and beliefs that were comprised in eugenics was sustained in eugenics societies.[16] Interestingly, eugenics societies did not disappear at this time. They merely changed their names to titles that sounded politically neutral (for example, the American Eugenics Society became the Society for the Study of Social Biology). Policies advanced by these societies, such as the forced sterilization of "feeble-minded" women, continued overtly until the 1970s, and continue today covertly in some state institutions.[17]

The particulate notion of nature versus nurture was also sustained in the new field of medical genetics. When medical genetics entered the scene soon after World War II ended, scientists began treating *diseases* like they were singular traits. They began the hunt for the "gene for" specific diseases. And they gave off the major misconception that if a mutation was found it was entirely responsible for the disease.[18]

At this time, nature versus nurture also came to be confounded with the many scientific and popular misunderstandings around the term "heritable." "Heritable" was originally used to mean anything passed down across generations. But in science, that meaning came to be conflated with the very technical, statistical term "heritability." "Heritability" refers to that idea originally issued by Karl Pearson—the percentage of variance in a population due to genes. As such, it's a population genetic analytic. There is no way to determine the extent to which genes are responsible for a particular trait in a particular individual. Nevertheless, scientists, journalists, and all sorts of people continue to use "heritable" and "heritability" interchangeably, in ways that give the impression that individuals can be assessed.[19]

It was amid these trends in the postwar years that behavior genetics established itself as a field in its own right, further perpetuating these misconceptions. Behavior genetics sought to study the impact of an organism's genetics on its behavior. It also was interested in the ways that heredity and environment interacted. As sociologist and historiographer Aaron Panofsky has shown, even though the ghost of eugenics haunted the study of behavior, animal behaviorists began to resuscitate the topic, creating a "disciplinarily and intellectually diverse research network throughout the 1950s and 1960s."[20]

Initially the field availed itself of eugenics, controversial racial comparisons, and bold statements about human nature. It also barred contentious scientists with reputations for questionable research interests and claims, especially around humans. The overall trend around the world was toward a fierce environmentalist backlash to eugenics, and the twin studies approach that the field believed was the only hope for a human-focused behavior genetics was coming under intense criticism. Critics argued that parents might be biased to raise identical twins more similarly than fraternal twins.[21] They also argued that twin studies held pernicious assumptions, such as the assumption that people really mate randomly as opposed to mating with people who share similar traits, or that environmental factors are less significant than genetic ones.[22] Though this criticism put behavior geneticists on their guard, none of these debates addressed the basic assumption growing dominant in science and the public that there was a disjoint between nature and nurture, between genes and environment.

In 1969, eugenics and behavior genetics took a near-fatal blow when psychologist Arthur Jensen wrote an article claiming "that genetic differences explain the IQ gap between blacks and whites, and educational efforts to close cognitive gaps were doomed to fail."[23] Though critics responded with an uproar, charging behavior genetics with racism, classism, and genetic determinism, they did not interrogate the particulate notion of nature versus nurture at the basis of the science. Controversy further kicked up as Stanford physicist William Shockley, a mentor to Jensen and a main supporter of his research, waged a campaign to reinstate sterilization to eradicate low IQ. Shockley petitioned for experts and members of government to reconsider Nazi programs in order to fashion a more "ethical" voluntary eugenic program that could target blacks. He was resoundingly denounced by an array of geneticists and behaviorists. At this time, eugenicist Cyril Burt died, and an investigation into scientific fraud on his part began. Burt, who had raised suspicion when he had purported to have found the exact same statistical disparity across a range of studies, had burned all his papers just before dying. This torrent of scandal forced behavior genetics to do some soul-searching, recoup losses, and reinvent itself.[24]

Throughout the 1970s, the behavior genetics controversy dominated the mass media as well as science. The population geneticists who ruled the behavior genetics opposition argued that nature and nurture should be completely separated, that race and biology should be redefined in stark zoological terms, and that behavior belonged outside the halls of genetics.[25] Behavior geneticists of all ilk debated whether and how to defend the field's stance that behavior was also genetic, and whether to excommunicate Jensen and the other rogue scientists, or find a way to repackage their findings as good science. Nearing the end of the decade, the field decided on the latter. Yet though they promoted a unified front, the field no longer could maintain its reputation as a thriving "inter" discipline. At this time, many behavior geneticists went underground— but not before branding in the public mind that even in behavior, genes were firmly opposed to the environment.

Sociobiology and the Crystallization of a "Nature First" Orientation

At the height of the IQ debates, as genetic notions of nature versus nurture were growing in public prominence, a group of population geneticists split off from the mainstream to make a new evolution-based science of behavior called "sociobiology." Sociobiology was based on the premise that social behavior, just like physical traits, resulted from evolution. The term "sociobiology" was actually first introduced by biologist E. O. Wilson at a 1946 conference on genetics and social behavior. But it didn't gain wide recognition until he published the book *Sociobiology: The New Synthesis* in 1975. In *The New Synthesis*, Wilson defined sociobiology as "the extension of population biology and evolutionary theory to social organization."[26] The first and last chapters of the book addressed human behavior in particular, and established a methodology of using hunter-gatherers in Africa (the San) as proxies for ancient peoples undergoing evolution. In 1978, Wilson published *On Human Nature*,[27] which squarely focused on the more controversial subject of human nature.[28] Following this book, a rash of evolutionary scientists turned to comparing human and animal behaviors like altruism and aggression.[29]

Like behavior genetics, sociobiology was heavily criticized for genetic determinism and for dismissing environmental effects.[30] In fact, the major critics of behavior genetics were the loudest voices against sociobiology, population geneticists who demanded a stricter zoological portrayal of human nature. In particular Harvard scientists Richard Lewontin and Stephen Jay Gould aimed two central criticisms. The first was against what they called "adaptationism."[31] They argued that sociobiologists assumed that behavioral traits were undergoing natural selection without proving it. Furthermore, sociobiologists didn't consider other evolutionary explanations for the existence of traits, such as chance or genetic drift. And because sociobiologists dealt in evolutionary theory, they didn't concern themselves enough with the "phylogenetic" (aka species-specific) or developmental particularities of different species. They just transposed one species' characteristics onto another willy-nilly and drew up theories therefrom.

The second central criticism regarded the sociobiological assumption that behavioral traits were heritable.[32] Critics like Lewontin and Gould pointed out that human behavioral traits were quite variable across geographical and cultural environments, whereas human genetic variation was not commensurate. With these ratios at such a mismatch, human behavior would never fit with the standing heritability model noted above, that variation could be accounted for primarily by genes rather than environment.

Added to this was the nagging problem of sociobiology's implications for race. As public health historian Michael Yudell writes, some sociobiologists came to assert that xenophobia was just a figment of the inborn aggression that humans possessed.[33] This focus on the biological mechanisms of racism became a basis for the rationalization of racism, henceforth serving "the needs of those who harbored racist ideas by giving them scientific legitimacy."[34]

But since sociobiology never formed a field as behavior genetics did, controversy over race did not force the science to stake a specific claim about race. Rather, sociobiologists were free to individually debate their theories about genetic causes, which along with their critics assumed a particulate version of nature versus nurture.

Despite the criticism, sociobiology found its way into "the core research and curriculum of virtually all biology departments," becoming "a foundation of the work of almost all field biologists."[36] It also birthed related fields in the social sciences, including human ecology and evolutionary psychology.[36] Yet perhaps its most influential contribution to the nature-nurture debates was its inspiration for a new theory of genetics and behavior encapsulated in *The Bell Curve: Intelligence and Class Structure in American Life*.

American psychologist Richard Herrnstein and American political scientist Charles Murray published *The Bell Curve* in 1994.[37] The book's central aim was to prove that "US class structure can mostly be attributed to inequalities in individual intelligence as measured by IQ, that IQ is . . . under genetic control, and therefore differences in education and upbringing are not responsible for social inequalities."[38] The book also claimed that there were racial differences in intelligence, and that these differences were due to genetics.[39] At the heart of Herrnstein and Murray's argument was the belief that intelligence was heritable in the particulate sense of nature versus nurture, and that nature was the fundamental cause of all things behavioral. The authors vociferously argued for the US government to get its head out of the sand and create policy based on these fundamental truths that nature provided.[40] In particular, they claimed that the

average genetic intelligence of the United States was declining due to three factors: the tendency of the more intelligent—or "cognitive elite"—to have fewer
children than the less intelligent; the shorter generation length of the less intelligent; and the large-scale immigration of people with "low intelligence" to the
United States. Short of arguing for eugenics, they designed a policy campaign
that would prevent immigration, affirmative action, and welfare. In fact, they
cast welfare as a troubling eugenic policy framework "that inadvertently social-
engineer[s] who has babies" and that was "encouraging the wrong women."[41]

The Bell Curve met a furious response across American science, garnering
international notoriety as well. It graced the covers of Newsweek, the New York
Times Magazine, and the New Republic, and it was featured on Nightline, the
MacNeil/Lehrer NewsHour, Charlie Rose, Primetime Live, and the McLaugh-
lin Group, where a wide array of experts charged Herrnstein and Murray with
an all-out revival of eugenics.[42] Yet even this strong response did not unseat
the still dominant view that nature and nurture were separate entities. Critics
merely argued that human cognitive ability couldn't be approached as a singular entity (especially not as a single number that could be rank-ordered in a
linear fashion),[43] that intelligence changed over the course of a life span with
respect to a person's environment, and that the IQ tests that Herrnstein and
Murray drew on were inaccurate and biased with regard to race, ethnicity, and
socioeconomic status. These criticisms drew on a strict environmentalist rationale that merely turned the tables on nature versus nurture. The environmentalist critique alleged that nurture, not nature, was the causal basis of behavior
and the racial disparities that The Bell Curve featured.

The environmentalist critique popularized in the media also did nothing to
break down the nature-nurture dichotomy for another reason. It simply didn't
dispel the "nature first" mantra within the natural sciences. By this point in the
mid-1990s, genetic science was resolutely headed in the direction of genomics—the study of DNA sequences. Sociobiologists were thrilled at the chance
to finally be able to back up their theories with DNA evidence, and behavior
geneticists, some of whom were finally considering turning toward a "holist
and environmentalist" approach,[44] heartened that finally they would be able
to back up their extrapolations to humans, to put their work on firm scientific
ground.[45] And though some scientists expressed hopes for a gene-environment
science somewhere in the distant future, the entire science community was en-
grossed with sequencing the DNA of different organisms, including humans.
Even "functional" genomics, or genomics that said something about how genes

worked, was but a distant dream. Instead "structural" genomics—what DNA looked like and how it was linked together—was the science of the day.

Almost immediate to the release of *The Bell Curve*, the American Psychological Association's Board of Scientific Affairs summoned a task force to investigate the book's research. While the task force issued a statement against its racial interpretations, declaring that there was "certainly no such support for a genetic interpretation . . . that the Black/White differential in psychometric intelligence [was] partly due to genetic differences,"[46] it came out supporting much of the science in the book. That same year, psychologist J. Philippe Rushton published *Race, Evolution, and Behavior*, a sociobiological treatise on the innateness of race and behavior, and then-president of the Behavior Genetics Association, Glade Whitney, called for his field to once and for all determine what the genetic roots of racial behavioral difference really were. As Panofsky notes, overall the "broadest, most public response was to embrace the arguments of *The Bell Curve*."[47] The only scientists who denounced the book's science were a small group of animal geneticists. Given that their research was not anywhere near the topics they were criticizing, their voice was drowned out by the majority vote of confidence. As journalist Jim Naureckas noted in his analysis of the Bell Curve media storm, "While many of these discussions included sharp criticisms of the book, media accounts showed a disturbing tendency to accept Murray and Herrnstein's premises and evidence even while debating their conclusions."[48] Critics were often painted in the media as experts who were simply possessed by political correctness.[49] All in all, *The Bell Curve* scored one for nature and a windfall for the nature-versus-nurture approach.

Evo Psych, Evo Cognitive Genetics, and the Development of Neuro (Social) Science

Sociobiology took a hit during the *Bell Curve* debate. But like eugenics a half century before, the science did not disappear. Instead, it shape-shifted into a new science called evolutionary psychology.[50] Evolutionary psychology examines behavior from an evolutionary perspective, but instead of theorizing about reproductive fitness by characterizing natural selection of traits across species, it sets its sights on the upstream causal processes of human behavioral traits like cognitive mechanisms of language, intelligence, and decision making.[51] Evolutionary psychologists are interested in what makes humans, and humans alone, tick. The field uses a combination of experimental methods, surveys, observation, and biometrics such as functional MRI (fMRI) to characterize innate drives for "face recognition, spatial relations, rigid object mechanics, tool-use, fear, social exchange, emotion perception, kin-oriented motivation, effort allocation and recalibration, child care, social inference, sexual attraction, semantic inference, friendship, grammar acquisition, communications-pragmatics, and theory of mind" as modules of adapted cognition.[52] In other words, evolutionary psychologists empirically assess human thinking, not just theorizing the outcomes of thought witnessed in the "phenotype" of behavior, and they see thinking as comprising a bunch of modular instincts. Because they start with the premise that these instincts evolved in our species over fifty thousand years ago, they study the behavior of hunter-gatherers today.

Though the field aims to tell us about the intersection of nature and nurture, it has remained as "nature first" as any of its predecessors. This predisposition is encapsulated in the title of one of the field's inauguratory essays, "The Psychological Foundations of Culture."[53] It is also evident in the language used by its key proponents, such as Harvard psychologist Steven Pinker, author of the Pulitzer Prize-finalist best seller *The Blank Slate: The Modern Denial of Human Nature*, who refers to behavior as controlled by a "built-in machinery" that makes up an essential human nature.

Evolutionary psychology has drawn many of the same criticisms that sociobiology has. It too is criticized for adaptationism, or assuming that all behavior has been selected for evolutionary purposes, and for its unsubstantiated emphasis on genetics.[54] But with its empirical claims connecting modern-day behavior to ancient human behavior, evolutionary psychology is also criticized for its unique brand of hypothesis testing. Critics argue that researchers know little about the environment in which humans evolved, so explaining psychological traits as adaptations to that unknown environment can only be mere speculation.[55] Critics also take issue with using modern-day hunter-gatherers as a proxy for ancient humans in staking empirical claims. As writer and broadcaster Kenan Malik argues:

> Behaviour, unlike bone, does not fossilize. Precisely because these are ancient humans, living before the development of much technology, archaeological evidence is scant. And there are certainly no ancient humans still living for us to observe . . . why should we assume that the lives of "primitive" societies today resemble those of ancient humans? Both may be hunter-gatherers, but the !Kung San, the Ache and others, are likely to have changed and developed over the past fifty or hundred thousand years. After all, no one would claim that modern

agricultural societies resemble those of the earliest farmers ten thousand years ago. Why make the same assumption about hunter-gatherers, especially over a much greater time span?[56]

Malik answers his question with the finding that evolutionary psychology is rife with analogy and metaphor, which its proponents mistake for literal reality. Instead of depicting what is measurable or better yet, what has been measured, evolutionary psychology weaves tall tales without proof.

This overreliance on figurative reasoning makes for a science that works backward from a conclusion in a circular logic. As Malik aptly asks:

> How do they know that ancient humans possessed those modules? They don't. They describe, rather, modern behaviours, needs and knowledge and then assume that ancient humans can also be understood so. Far from using ancient behaviour to understand modern behaviour, evolutionary psychologists use modern behavioural categories to reconstruct our ancestors' minds.[57]

Thus, modern racism is translated as an inborn self-preserving categorization system, and modern sexism (including child-rearing disparities, the glass ceiling, and even rape) is just our species sorting roles according to our natural physiological strengths and weaknesses.

Beyond its scientific inadequacy, critics have derided evolutionary psychology for lending itself to the justification of existing social hierarchies and reactionary policies. Indeed, its claims have inspired hate groups such as the neo-Nazi group "Stormfront" and the Ku Klux Klan. In 2011, evolutionary psychology suffered from its own race scandal when London School of Economics psychologist Satoshi Kanazawa published "A Look at the Hard Truths about Human Nature," a *Psychology Today* blog post in which he popularized the substance of his work "Why Are Black Women Less Physically Attractive than Other Women?" and "Why Men Gamble and Women Buy Shoes: How Evolution Shaped the Way We Behave."[58] Kanazawa warned that neither the higher "mean body-mass index (BMI)" of black women nor "the race difference in intelligence (and the positive association between intelligence and physical attractiveness)" could account for the purported disparity in attractiveness: "Black women are still less physically attractive than nonblack women net of BMI and intelligence. Net of intelligence, black men are significantly more physically attractive than nonblack men."[59] Kanazawa guessed that an inordinate amount of inborn testosterone, evolved from an ancient racial divergence, was the culprit.

As with *The Bell Curve*, this treatise kicked up a great deal of public controversy, while still producing a mixed response from within the sciences. Following outraged responses, *Psychology Today* changed the article's title to "Why Are Black Women Rated Less Physically Attractive than Other Women?" and then took it down altogether.[60] *Psychology Today* then wrote an apology:

> Last week, a blog post about race and appearance by Satoshi Kanazawa was published—and promptly removed—from this site. We deeply apologize for the pain and offense that this post caused. *Psychology Today*'s mission is to inform the public, not to provide a platform for inflammatory and offensive material. *Psychology Today* does not tolerate racism or prejudice of any sort. The post was not approved by *Psychology Today*, but we take full responsibility for its publication on our site. We have taken measures to ensure that such an incident does not occur again. Again, we are deeply sorry for the hurt that this post caused.[61]

Students at Kanazawa's university, the London School of Economics, called for his dismissal,[62] and the University of London Union Senate, a union representing over 120,000 students, voted unanimously for the same fate.[63] While a cohort of sixty-eight evolutionary psychologists issued an open letter distancing the field from Kanazawa's claims, another open letter by an international group of twenty-three scientists defended them.[64] Again, critique was aimed overall at the racism inherent in Kanazawa's claims, but not at the fundamental assumptions of evolutionary psychology. And no attention was paid to the nature-nurture dichotomy that his research posed.

However extreme these views seem, similar arguments have been increasingly at the forefront of the emerging field of evolutionary cognitive genetics, the first field completely dedicated to mapping brain genes. Panofksy traces the influx of molecular geneticists into the field of behavior genetics in the mid-1990s; rather than acting "like sober, objective experts," they made "wildly speculative, determinist, and 'astrological' claims about reading individuals' fates from their genes."[65] They forged the path for a new interdiscipline, which historian of science Sarah Richardson describes as "best characterized as a convergence of research trends in genomics and the brain sciences, represented by an informal and diverse constellation of brain genomics, human population genetics, and molecular anthropology researchers."[66]

Evolutionary cognitive genetics is really a proto-sociogenomic science, in that it studies the genomics of mental and intellectual health and disease using actual DNA sequences pertaining to human behavior. From this new orientation, scientists have broken ground on a number of topics in neuroscience and population biology, including the controversial issues of intelligence and race. Richardson points out that evolutionary cognitive geneticists believe that the dozens of genes implicated in evolutionary cognitive genetics studies "will surely yield examples of genes with variants of different frequency in different populations, portending a wide-open pathway for genetic claims about racial or ethnic differences in brain and cognition." Some even herald a future of race-based medicine, in which drugs can be targeted to correct IQ disparities in African Americans.[67]

Like the previously described sciences, evolutionary cognitive genetics starts from the premise that natural selection is operant in all human behavior. And like the other sciences discussed, it uses narrative to spin stories of what might have happened in the past. For example, Richardson finds more than one case where evolutionary cognitive geneticists have recounted that the evolution of the human brain to its modern status happened only after humans migrated out of Africa.[68] She also shows how notable members of the field who work on interspecies comparisons shuttle between stories of animal behavior, ancient behavior, and the habits of groups residing in Africa to embellish their stories.

This science has received a great deal less criticism than other fields, most likely because it is largely focused on characterizing actual DNA. While specific narrative statements, such as the non-African human intelligence departure point tale, have been lambasted in the science media, there have been no public scandals to date. As such, evolutionary cognitive genetics' "nature first" assumptions have not been scrutinized or challenged.

Moving away from the focus on intelligence and basic cognition, there are several other fields that have straddled the fence between biology and psychology, inching toward the sociogenomic. The field of neuroeconomics examines the physiological origins of human decision making.[69] It is an offshoot of behavior economics, or microeconomics, that is interested in complexifying the standard rational choice and game theories of economics that assume that humans make rational, self-interested decisions in order to maximize their wealth and well-being. As economists Nick Wilkinson and Matthias Klais put it, "instead of hypothesising about Homo economicus" the field's founders decided to base research on "what actually goes on inside the head of *Homo sapiens*."[70] With a battery of new psychological insights into how humans construct decision-making models, and an interest in the deeper chemical processes active in the human brain, founders incorporated methods and approaches from theoretical biology, computer science, and mathematics.

Yet it wasn't until brain-imaging technologies became available that the field began to institutionalize.[71] As the founding president of the Society for Neuroeconomics, Paul Glichmer, and economist Ernst Fehr summarize, looking back at their field's initial formation in the early 2000s:

> A group of behavioral economists and cognitive psychologists looked towards functional brain imaging as a tool to both test and develop alternatives to neo-classical/revealed preference theories . . . [and another] group of physiologists and cognitive neuroscientists looked towards economic theory as a tool to test and develop algorithmic models of the neural hardware for choice. . . . The result is that the two communities, one predominantly (although certainly not exclusively) neuroscientific and the other predominantly (although not exclusively) behavioral economic, thus came towards a union from two very different directions. Both, however, promoted an approach that was controversial within their parent disciplines.[72]

Functional MRI moved behavioral economists' empirical emphasis from the survey end of their research to the physiological end, thus pivoting the field from an environmentalist, nurture-interested science to a nature-focused one.

Though the field has not made the "nature first" or "nature only" statements that prior fields have made, it has provoked criticism for its implication that nature is the fundamental cause of human behavior.[73] For example, in "The Case for Mindless Economics," economists Faruk Gul and Wolfgang Pesendorfer (2008) argued that reductionistic approaches that may have been appropriate and successful in the natural sciences are "unlikely to be able to relate natural scientific phenomena to social scientific theory."[74] Because economic theory makes predictions about behavior, it must thus be skeptical about the actual mechanisms behind human decision making.

Similar to neuroeconomics, neuropolitics combines neuroimaging methodologies with theoretical frameworks from neuroscience, political science, psychology, behavior genetics, primatology, cognitive science, and ethology in order to investigate the relationship between the brain and politics. Researchers study classic questions from political science such as how people form political attitudes and make political decisions, as well as how the evolution of political competition has affected brain development in humans and other species. As with neuroeconomics, neuropolitics launched with a focus on behavioral experiments but seized the opportunity to apply fMRI to retrain its sights on neurological processes. A typical neuropolitics study involves presenting subjects with various political scenarios and then mapping reception in the brain.[75] Studies also often focus on comparing differences in liberals and conservatives.

Like neuroeconomics, the field has provoked criticism for implying nature to be the main cause of behavior. Critics have censured neuropolitics, stating:

> In the rush to neuroscientific discovery important questions are overlooked, such as the ways: (1) the brain, environment and behavior are related; (2) biological changes are mediated by social organization; (3) institutional bias in the application of technical procedures ignores race, class and gender dimensions of society; (4) knowledge is used to the advantage of the powerful; and (5) its applications may reinforce existing structures of power that pose ethical questions about distributive justice.[76]

Critics highlight the ways in which neuropolitics exacerbates the knotty issues of ethics in a science that presumes a nature-nurture split, and they entreat scientists to mind how their research will be taken by the public.

One last field that I draw your attention to is the field of neurocriminology. Neurocriminology draws on diverse scientific tools developed in fields like evolutionary psychology and neuroscience to establish neurological causal mechanisms for crime. It is a subfield of biocriminology—the science of biological determinants of criminality.[77]

Biocriminology itself is a field over a century old. It emerged in the late nineteenth century with Cesare Lombroso's Italian School of Criminology. Lombroso was an army physician who argued for the heritability and incurability of criminality.[78] Lombroso claimed that criminals were "evolutionary throwbacks who could be identified by 'atavistic stigmata' such as beaked noses, fleshy lips, and shifty eyes: features suspiciously suggestive of race."[79] Lombroso's evolutionary theory was premised on a racist theory that assumed that humans were a spectrum of development, wherein Africans were most primitive and Europeans most evolved.[80] While Lombroso was alive, his theories supported the rapidly popularizing eugenic sciences and policies.

What is important about the early work of Lombroso and nineteenth-century biocriminologists is that they established a method that continues to operate in the field of neurocriminology today, in which scientists scour criminals for markers of genetic deficiency. Lombroso's first biocriminological study involved performing an autopsy on a serial killer. In the killer's brain and nervous system Lombroso observed atrophy, which he claimed was inborn.[81] Throughout his life, Lombroso would replicate this approach in different populations considered to be "feeble-minded," such as mentally ill and retarded people, also targeting people who displayed left-handedness or tattoos.[82] His science and the biocriminology of the twentieth century was not only a supporter and pillar of the particulate notion of nature and nurture, but also a central part of the institutionalization of methodologies that created the population statistics, evolutionary science, and medical genetics of the twentieth century.

Trends around explanations of crime have historically swung like a pendulum between sociological and biological theorizing.[83] Of late, the field has swung in the direction of biological explanations and genetics. As criminologist Nicole Rafter argues in her masterful history of biocriminology, *The Criminal Brain*, biocriminology is enjoying a resurgence because of the broader societal turn to genetics and also a growing desire to prevent harm by any means necessary.[84] Thus the swing toward neurocriminology means a swing toward natural causes and the potential manipulation of nature to change nurture.

Neurocriminology uses many of the same terms as evolutionary psychology, neuroeconomics, and neuropolitics, such as "drives," "risky behavior," and "extraversion,"[85] and it relies heavily on brain imaging. However, the field has an added social dimension in that it is used in criminal litigation. Neurocriminologists have been called on to provide evidence for everything from hormonal imbalances and PMS to innate violent tendencies.[86] Thus, in addition to garnering criticism for biological determinism, it is also criticized on the point that it disallows for rehabilitation.[87] Its fervent belief in the biological origin of human social behavior, and its drawing of the most distinct of lines in the sand between nature and nurture, has wider social implications that expose the policy dangers of "nature over nurture" science.

These fields can be thought of as the primordial soup for sociogenomics. Their common denominator is that they all privilege biological, if not genetic, explanations for social phenomena, and they uphold the myth that nature and nurture are ultimately distinct. Although scientists in these realms study the relationship between social measures and biological ones, they maintain the "mirage of a space between nature and nurture" that Keller so vividly described in her eponymous book.

But before we move on, I want to emphasize that even the most mainstream social sciences from which sciences like these have sprung (economics, political science, sociology), which have elaborated alternative nurture-based explanations for generations, have for centuries been riddled by debates that have left an aperture for interest in emerging genetic models. From their start, these sciences have wrestled with questions of whether it is macro structures like government and education that make people and societies what they are or whether micro experiences of day-to-day interaction are more influential. Similarly, they have debated whether social structure or human agency is a better explanation for human behavior. Is it more important to focus on market fluctuations or consumer behavior? Congressional redistricting or community volunteerism? School policy or parenting? These central questions do not obviate the incorporation of genetic explanations for who we are; rather they sidestep direct engagement with the question of the role of genetics. There is nothing inherent in these competing paradigms to warrant an all-out rejection of a gene-environment approach.

Moreover, the core of these social science disciplines has not been as opposed to natural scientific explanatory frameworks as one might assume. My own field of sociology is a typical example. Its founders envisioned the field to be the last stage in an evolution of the sciences. Sociology would, just as physics and biology had, reveal the universal laws of our world. Indeed, sociology borrowed much of its foundational language from these natural science fields. Terms like density, differentiation, diffusion, and volume were applied to society wholesale, becoming "social" density, differentiation, diffusion, and volume. Founding theorists discussed the organism of society or the "body politic" and its component parts in the same way that biologists discussed the human body. Even as the field developed its emphasis on cultural beliefs and interpersonal interaction, the notion that genetics could play a role in human behavior was not dispelled. And as sociologist Bernice Pescosolido has noted, the few times that the field has assessed its "internal barriers to progress," elites have merely argued for better knowledge of biology and scientific data through participation.[88]

The now popular social science tenet that the world around us is socially constructed, that is, made by our own hands on the basis of shared views and understandings, continues to appear together with seemingly contradictory determinisms like genetic determinism, even in areas where one would think one or the other would prevail. As such, sophisticated gene-environment models have enticed social sciences working at the heart of these disciplines, adding one more resource in their explanatory armamentarium that can get at the fundamental cause of why we do what we do.

Human Genomics and the Launch of Sociogenomic Science

Right about now you might be wondering where good old-fashioned genetics is in all this? When in the 1990s genetics was converted into genomics—the study of gene sequences and DNA throughout the genome—science headed in a decidedly biomedical direction. As mentioned earlier, genome scientists initially did not work on behavior. Instead, they spent their energy solving technical issues around sequencing human DNA and mapping the genomes of an array of organisms. Of the genome leaders that I wrote about in *Race Decoded*, many were strictly opposed to a field of behavior genomics well into the late 2000s.[89] Some had downright harsh words to say about the application of genomics to human behavior. The general vibe was that genomics should be about preventing illness, not determining cognitive prowess.

But with the successful mapping of the human genome and the landslide of innovations in sequencing whole genomes of individuals, all that changed. As basic genomic techniques were improved to work on greater regions of the genome in greater detail, and as genetic pathways (including how genes actually work in specific organs and cell types) were discovered, a gene-environment initiative overtook public health and biomedical establishments across the world. This can be seen in the burgeoning world of toxicological genomics and the series of newly formed "omic" sciences, like the "exposome" and the "microbiome." Sciences like these analyze the interaction between DNA and the environment of food, air, and water.

Also, there is the new science of epigenetics, which studies the DNA strands that surround genes on our chromosomes. These outside strands regulate whether specific genes are turned on or off in certain environments, as well as over the life course. Epigenetics has found that everything from diet and exercise to parenting strategies and cultural conventions affect whether the chemical pathways to turning genes on and off are sparked.

Through these initiatives, and through a growing interest in the study of diseases like Alzheimer's and schizophrenia, the leading genome research institutes have found their way to studying human behavior. Now many have a branch devoted to studying behavior, if only from the perspective of mental health and illness. Increasingly, it is publications of these behavioral studies that grace the headlines, bringing notoriety to institutes and the field.

Many mainstream genome scientists are now embarking on debates about the possibility of innate behavioral tendencies. Take Francis Collins, the current director of the NIH. In many venues Collins has expressed his belief in an innate moral orientation.[90] Darwin also believed in this. But unlike Darwin, Collins's views sit in the context of a swarm of studies that argue for the neurological ubiquity of human morality. Scientists like him believe that the genomic mechanisms behind our basic social orientation to the world can and will be revealed.

Indeed it is mainstream public health and personalized medicine interests that are leading the way in innovation around behavioral testing and modification. In the summer of 2009, a team of scientists at Shanghai Biochip Corporation and educators at Chongqing Children's Palace administered genomic talent tests to children 3–12 years old at a "genetics summer camp." Tests covered a spate of behavioral traits not limited to IQ, memory, focus, emotional control, and athletic ability. Predicting sports and musical ability were two key areas of interest. But the lead scientist on the program, Huang Xinhua, said that the team was also dedicated to predicting future careers. In an interview with CNN, as he looked over one student's test results, he exclaimed: "This child is very thoughtful and focused, so I suggest she go into management." The team projected the viability for testing in all Chinese families, especially given China's One-Child Policy and the desire it created in parents to maximize their child's education.[91]

In the spring of 2015, scientists at Sun Yat-Sen University in Guangzhou, China, made the first edits to the human germ line.[92] Though their work was aimed at a medical disorder—beta thalassemia, a group of blood disorders that most Americans know by the disease sickle cell anemia—it was conducted to see if it would be possible to edit out undesirable traits before conception. The experiment was successful in the editing process, but the scientists were unable to produce an embryo that didn't contain collateral damage spread throughout its genome. Still, they expressed hope that future attempts would produce better results, such that one day a complete preimplantation genomics of a range of desirable traits could be a reality. By December 2015, governmental health institutes from the United States, United Kingdom, and China were all meeting to decide on how to proceed with innovation around behavioral and cognitive traits.

Genomics has thus, with its full-on bridging of the social and genomic, given birth to sociogenomic science. It is its very own gene-environment initiative that has laid the foundation for the paradigm and that has made emerging fields like social genomics, the focus of much of this book, so quickly take root. It has created a form of science that goes beyond theory to build claims around the G's, T's, C's, and A's of our human nucleic code.

How does sociogenomic science hold up in terms of nature-nurture conceptions? As I hinted in the Introduction, this novel approach to gene-environment interactions does not preclude a conceptual split between nature and nurture. It doesn't mean that experts are really thinking outside the binary of nature versus nurture, and it definitely doesn't imply that nurture will be taken seriously. The contemporary genomic way of approaching gene-environment science and human behavior is still "nature first." It's obvious that studies that only characterize DNA sequences, ones that only use social survey research to sort cases and controls, are not doing due diligence with the study of the environment. All this leaves us in the same morass that we were in at the advent of modern science. Today's world is still plagued by the polysemy of early biology.[93] Now more than ever our "gene" concept needs a retrofit.

Keller offers a solution. She calls for scientists to take the focus off of the word "gene" and replace it with the language of "DNA, RNA, proteins and pathways, and chemical inputs from our being in the world." She says it's better to think of DNA in terms of a resource among others on which a cell can draw to respond to its environment. And instead of focusing on heritability, she wants us to think in terms of phenotypic plasticity, or how malleable a given trait is, and when in the life course it changes ("how malleable individual human development is, and at what developmental age"). We normally set birth as the cutoff point between nature and nurture, but everything from height to linguistic skill is changing in us often past adolescence and into our adult lives. Trait development happens before, during, and often long after birth, so taking a long view of human being (should we call it "nature/nurture," or maybe "narture"?) will help to dispel the ignorance and confusion that permeates our collective understanding.[94]

Sociogenomics' Meaning for Politics Today

The presence of sociogenomics has the potential to radically impact the ways that humans socialize and are socialized, how we politicize and are politicized. Before I turn to the next chapter, revisiting thoughts raised in Race Decoded,[95] I want to pause for a moment to consider whether sociogenomics is indicative of a new form of sociality.

More than one contemporary observer has argued that the new genomics, with its technologies and techniques of redefining one's understanding of oneself, is creating a new world. Anthropologist Paul Rabinow coined the term "biosociality" to talk about how genomics brings with it a totally new way of relating, focused on mutability and optimization. According to Rabinow, sociobiology and the nature-versus-nurture sciences before were essentializing and normalizing—in other words, they reduced everything to the essence of DNA and made it seem like there was a species-wide "normal" that all others would be held up to.[96] But genomics carries a new orientation to life where the focus is now on performance relative to one's personal biology or, as anthropologist Kaushik Sunder Rajan puts it, "risk and difference, rather than deviance."[97] DNA is no life sentence.

In this brave new world, old ways of grouping such as by race or gender might still matter, but they will over time be rewritten by new biosocial categories. People will care more about their genotype at a particular locus than their phenotype, or what they look like on the outside. They will prioritize socializing with people on the basis of shared innate traits, including collectively organizing for political rights and resources based on such traits.

Philosopher of science Ian Hacking agrees.[98] Before, kinship was the operative grouping. But now anything goes. More and more social groups are characterized by a combination of biological and social qualities, and not in some deterministic, prefab way. Genomics creates *more* choices for people's making of themselves. Meanwhile, society is rapidly evolving to meet the technological cornucopia of tests and drugs, finding new applications all the time. Passage points for the "making up" of people, such as immigration, targeted medicine, slavery history recuperation, and indigenous political rights and reparation are being redefined in genomic terms.

Sociologist Nikolas Rose sums this view up with the phrase "biology is opportunity."[99] A shifting of concern with exhaustive biological morphology toward minute snippets of our DNA at the molecular level creates a new politics where optimization is the logic. Managing one's genomics vis-à-vis their myriad interactions with the environment (eating breakfast, brushing teeth, walking or driving to work) is the collective socializing process. Opportunity brings hope. And the present is reconfigured as the future of individual potential and possibility.

But is the current moment really defined by a new form of biology-driven choice? I agree with these scholars that new biotechnologies are infiltrating our lives and radically reshaping our relationship to each other and ourselves. From birth, and often before, we are subjected to genetic screens. Even the luckiest children, those who never break a bone or find the need to visit a hospital, make preventative visits to the doctor in which their health is interpreted against a growing body of public health genomic literature. Drugs are aimed at their "molecular selves," or the specific cellular pathways tied to their genomes. Meanwhile aspects of their health data are fed into online systems that can be linked to their mobile devices and consulted at will (as I write this, I am signed into four daily health monitoring apps and two e-health systems that hold all my health records on my phone. This is not to mention the health app that comes embedded in my phone, which runs in the background day in, day out, and which can't be removed from it).

Still, the question remains of whether this is best described as opening up choice. In *Identically Different*, geneticist Tim Spector paints a picture of how gene-environment science will soon transform our lives. He writes, "In the future, when faced with a messy divorce, rather than resort to Prozac or gin and tonics we could take specific highly tailored chemicals to epigenetically steady our sensitive genes until the crisis has passed."[100] The idea is that sociogenomic therapies will direct us to a personal harmony between our bodies and the environment. Instead of suffering, we will reprogram ourselves to thrive.

This vision exposes one aspect of not so much a curtailing of freedom that comes with sociogenomics but an equally delimited situation as that of the past pregenomic era. Yes, being able to pop sociogenomically targeted drugs may liberate a person from stultifying depression or paralyzing rage, but it doesn't liberate people from the real social conditions that affect them. In fact, popping a pill will likely lead to pacification, and potentially depoliticization.

In my research on identity-based struggles to change racist and sexist policies—what many refer to as "identity politics"—I've found that the emerging forms of genomics do not preclude people from identifying with older phenotypic racial categories.[101] Rather, people are now using genetic tests to determine their race and then using test results to petition for things like affirmative action, tribal enrollment, and race-based healthcare. In a sense, race relations stand to continue unchanged in the presence of genomics: same categories, same meaning, same world of racial inequality that demands mass social action.

What's important here is that even though the new technology is being used to uphold prior categories that have been the basis of legacies of civil rights organizing and identity politics campaigns, and though many are using tests to garner political rights and resources, the test-taking process dispels grassroots politics. Test takers get their results and draw on them to substantiate official classification claims on their own, or they use one of the many chat rooms and social media outlets to confer with others who have had similar results about their common distant ancestry. Organizing around experiences of social injustice falls out of the picture. Mass social action is replaced by recreational mingling and learning about one's history through sociogenomic self-help literature.

Socializing is happening, yes, and it's increasingly becoming a biomedical matter; yet it is transforming the meaning of politics in ways that do not bespeak greater liberation. And it's in every drug and test maker's interest to keep the process that simple. Scientists promote an image of DNA self-analysis as fun. They spin tales of the true meaning of ancestral lineages with exciting stories of explorers and royals so that "root seekers" can revel in their literal birthright.[102]

A similar issue is raised by the vision often shared with me by social genomicists, that of a world in which policy is tailored to people's personal sociogenomics. The hope is for parents to be able to test their fetuses for characteristics that will aid them in providing the best environment for their child to grow in. This would be met with testing by educators and perhaps even employers at specific moments of development along the life course. Again, the goal is to nurture and track individuals for success given their unique DNA profile.

It is not hard to imagine how a world like this could usher in a new determinism that would not be liberatory by any means.[103] There are plenty of movies that show just that. Take *GATTACA*, the 1990s blockbuster that portrayed the future of prenatal DNA testing and gene therapy, where parents could design their fetuses to become superhuman. The world becomes a system of fixed hierarchy à la choices made. Is biology opportunity? Yes. But is it no longer destiny? No chance.

Let's reverse back to the present. How is sociogenomics creating a field of opportunity for making us up today? Is this nascent science and burgeoning media discourse presenting us with a new set of choices? I see sociogenomics bolstering deterministic ideas in society even as it generates a new ethos of hope for self-improvement simply because what it refers to—the DNA in our bones—is seen as such an essential, innate, and immutable reality. The reason that sociogenomics is taking hold so rapidly is that it presents knowledge that everyone cares about. Who doesn't want to know whether they are destined for educational success, mental stability, or lasting love? Who wouldn't want to optimize their chances for a life well lived by understanding what can't be changed? Sociogenomics brings a promise to people, but one that today still inherently depends on a nature-versus-nurture conception of life.[104] It encourages us to look within as opposed to all around us for change. Nurture can be bent and shaped to meet nature, but nature determines just why that will happen and how.

CHAPTER 2

SCIENCE WITHOUT BORDERS

There is no reason why social scientists should be left out of the gold
rush of analysis that is ensuing from the decoding of the human genome.
—Dalton Conley, coauthor of *The Genome Factor*[1]

The sociogenomic approach to understanding life, witnessed in the last decade of genomic research, is taking the world by storm. Now at the forefront is
the nascent science of social genomics, a science that has determined to focus
strictly on the relationships between the social and the genomic. But what exactly is social genomics? Who constitutes it, when and where is it practiced,
and why is it with us in this particular historical moment?

Here I map the landscape of social genomics, demonstrating the formation of a new autonomous scientific field. But social genomics isn't your average discipline-based field. From senior scientists' efforts to form nontraditional
laboratories and train doctoral students in genomic methods to the cohiring
of postdoctoral associates into medical and nonmedical departments and the
recruitment of junior faculty who are interested in a career centered on transdisciplinary innovation, I show that a unique form of field building is underway. This kind of field making is all about flexibility and openness and a kind
of nonspecific expertise that portends a vibrant growth in the years to come for
itself and the sociogenomic paradigm.

Social Genomics Gets Started

Social genomics has a fascinating life story, one that bespeaks the interdisciplinary nature of postgenomic science. Social genomics arose out of a series of
interdisciplinary career shifts and partnerships that started in the early 2000s,

when many leading social scientists took interest in the mapping of the human genome and its promise to revolutionize science. As one economist put it:

> I started to be interested in this basically around 2001 when the Human Genome Project was finished. Then, it sort of became obvious to me that maybe there is something in it that is also interesting for social scientists. But it really seemed to be very, very farfetched. I had absolutely no idea how you would possibly go about this, and then who you would need to talk to and who has data and how.[2]

Back then, there was only one social science cohort study doing anything with genetics. The National Longitudinal Study of Adolescent Health or "Add Health," a study of "a nationally representative sample of adolescents in grades 7–12 in the United States during the 1994–95 school year," had begun collecting data on a small sample of twins.[3] Social scientists were enabled to do heritability studies that compared the presence or absence of a certain trait in their twins sample by comparing the identical twins with the fraternal ones. Some genetic data were also collected on this subsample and made available to an elite group of social scientists during Wave III of the study.[4]

With Wave IV, collected from 2003 to 2008, Add Health started genotyping all of its participants for public use.[5] Other cohort studies, such as the Wisconsin Longitudinal Study,[6] the UK Biobank Project,[7] and the Health and Retirement Study,[8] followed suit. Meanwhile, various twins registries around the world and genomic databases opened their doors to social scientists.

A watershed moment came when an article by a team of behavior geneticists at Duke University published one of the first gene-environment findings based on the Dunedin cohort of New Zealand. Avshalom Caspi and Terrie Moffitt compared gene variants associated with the greater or lesser production of proteins associated with aggression (monoamine oxidase A, or simply "MAOA") in male participants, and found that the variant interacted with how much abuse participants had suffered as kids to predict their antisocial behavior as adults.[9] One year later, this team found a similar interaction between MAOA, stressful life events, and later-life depression, but in a mixed-sex population.[10] These studies shone a light on the potential for using longitudinal social survey research that marks the past and present in conjunction with genomics to make behavioral predictions for the future.

When this article came out, several sociologists were just embarking on a midcareer turn to genomic science. For Jason Boardman of the University of Colorado, it all began with a talk he gave that drew interest from sociologists working in public health. At the end of his talk, a VIP from the National Institute of Child Health and Human Development who was running the National Children's Study asked him to collaborate with their Social Environmental Working Group on neighborhood contexts. Through this collaboration, Boardman was recruited by a fellow sociologist for a National Institute of Child Health and Human Development Mentored Population Research Scientist Development K-Award.[11] This grant supported Boardman in going from knowing nothing about genomics to applying for and eventually running his own five-year NIH-funded lab in genomewide association studies, and to eventually establishing Integrating Genetics and Social Science (IGSS), a transdisciplinary social genomics research network. This start also led to Boardman's reception of the first-ever Population Association of America Early Career Award for "PAA members who have made distinguished contributions to population research during the first ten years after receipt of the PhD."[12]

At the same time, Dalton Conley, then at NYU, started doctoral studies in biology at its Center for Genomics and Systems Biology. During the course of his early years in the program, he was able to publish some of the first social genomics studies on interactions between genotype, family structure, and academic performance. He also established the field's first set of best practices around characterizing the environment. Conley modeled using "exogenous shocks"—things like wartime drafts and famines—as instrumental environmental variables. He would soon after take the position of dean of social sciences, through which he would recruit emerging social genomics scholars, such as David Cesarini and Christopher Dawes, to NYU's faculty.

Meanwhile, a group of political scientists at the University of Nebraska turned their political psychology lab into a gene-environment lab. Kevin Smith and John Hibbing both had spent a long career in policy, but something nagged at them. Their nonbiological models didn't seem to fit what they were seeing in their data. At a departmental lunch one day, Smith admitted to Hibbing that he was reading Darwin and Lamarck, going back to the source, so to speak. Within months, the two would launch a full research program in the area of political behavior genomics.

Rose McDermott, a political scientist then at UC Santa Barbara who had been interested in genetics for some time, also formed a relationship with the Nebraska group and a new political scientist on the scene, UC San Diego's James Fowler. At a 2005 conference at Rice University, in a session on the heritability of political preferences, they discussed the possibility of genes being responsible for the stable and heritable elements of political behavior. Fowler's research on cooperation, research that drew on behavior economic game models, was pointing to some kind of innate and biologically transferred human quality. He proposed doing a classic twin study to establish the heritability of political participation.

To accomplish the study, Fowler enlisted then-graduate student Christopher Dawes to attend the Annual Minnesota Twins Behavior Genetics Seminar and report back on their candidate gene approach. Eventually the two published two seminal papers, "Born to Lead: A Twin Design and Genetic Association Study of Leadership Role Occupancy"[13] and "Genetic Variation in Political Participation,"[14] on the Minnesota Twins data. McDermott and the Nebraska group also embarked on a research program focused on MAOA and male aggression.[15]

Several people who would go on to build the field out and publicize it across the sciences were able to take advantage of the interdisciplinary research opportunities already available at their home institutions. Sociologist Guang Guo of the University of North Carolina, the institution that houses Add Health, worked directly with the cohort study to run a number of studies on adolescent behaviors associated with promiscuity, binge drinking, and the like.[16] Sociologist Jeremy Freese, then at the University of Wisconsin, worked with the Wisconsin Longitudinal Study to revitalize his doctoral interest in melding sociology with behavior genetics.[17] These scientists found that their work on social genomics drew more of an interest than their other nongenetic research, and they were being repeatedly called to speak to power organizations like the American Sociological Association.

By the mid-2000s, doctoral students were also seizing the opportunity to do interdisciplinary work, immersing themselves in genomics and forging a transdisciplinary path. Policy analyst Jason Fletcher, who was an eventual cofounder of IGSS and coauthor of *The Genome Factor*, used the Robert Wood Johnson Health Policy Scholars forum at the University of Wisconsin to try out a talk on integrating genetic and social science data. The success of this talk spurred him to get his hands on the Add Health data. By 2005, the last year of his doctoral work, Fletcher was posing some of the field's seminal health economics questions. In 2006, when he took his first job at Yale, he launched a research program to continue using and building the use of genomic data in health economics and policy. Others had to move outside their home disciplines to make these career moves. Kevin Beaver, who was doing his PhD in sociology at the University of Cincinnati, started using Add Health data to understand genetic factors in criminal behavior. His doctoral research set the standard for criminology and earned him a place in that field, but it cost him the PhD in sociology. Pete Hatemi, a political science student at Nebraska, found himself flying halfway around the world to Australia to study with behavior genetics giant Nick Martin. With his unique blend of expertise, he would later become a key collaborator on an array of genopolitics studies, and the editor of its famous book, *Man Is by Nature a Political Animal.*[18]

In 2008, a wave of studies was released in the top journals of each of the social science disciplines. Fowler and Dawes published "Genetic Variation in Political Participation" in *American Political Science Review*, and in *Science.*[19] With his grad students Michael Roettger and Tianji Cai, Guo published "The Integration of Genetic Propensities into Social-Control Models of Delinquency and Violence among Male Youths" in the *American Sociological Review.*[20] Ariel Knafo and colleagues published the first candidate gene study in economics, which found an association between genes and generosity in students playing the "dictator game."[21] Even religion was coming into view with studies published by Duke sociologist Matthew Bradshaw and his grad student Christopher Ellison.[22]

At this point, two teams of economists began the first collaborations that would lead to the formation of the first social genomics consortium, the Social Science Genetic Association Consortium (SSGAC). One team, at Harvard, was led by economists David Laibson and Ed Glaeser. Laibson and Glaeser mentored grad students Dan Benjamin and Jon Beauchamp to conduct a series of studies using Iceland's Heart Association data.[23] The other team, at Erasmus University in the Netherlands, arose when new hire Philipp Koellinger began collaborating with Roy Thurik and Patrick Groenen to build ties with the Erasmus University Rotterdam Institute for Behavior and Biology, the Netherland's leader in genetics.[24] Their collective eagerness to use the center's cohort data to understand entrepreneurialism led to the launch of the first entrepreneurialism genomics studies and the Gentrepreneur Consortium.[25]

In 2011 this group convened at the Cohorts for Heart and Aging Research in Genomic Epidemiology Consortium (CHARGE) meeting in Boston to form a social behavioral working group.[26] Soon after, they began planning a piggyback social genomics consortium that would focus on gene-environment interactions in ways similar to CHARGE's big-data approach.

As the decade came to a close, social genomics researchers across these groups got multiyear conference grants from the NIH and NSF, as well as national organizations like the Population Association of America, the Russell

Sage Foundation, and the National Economic Bureau of Research, to form the IGSS and SSGAC. The first IGSS meeting, held in 2010, covered topics like "Genes, Economics, and Happiness" and "A Genome Wide Association Study of Educational Attainment."[27] The first SSGAC meeting, in 2011, similarly staked out a research program on educational attainment.[28] Both meetings sparked high-profile joint publications. IGSS compiled conference presentations into a special issue of *Biodemography*.[29] SSGAC published in the *Journal of Economic Perspectives* in 2011.[30] Importantly, both led to the establishment of social genomics' web presence, which secured the field's identity. The SSGAC's website developed their mission, stating:

> The SSGAC is a cooperative enterprise among medical researchers and social scientists that coordinates genetic association studies for social science outcomes and provides a platform for interdisciplinary collaboration and cross-fertilization of ideas. The SSGAC also tries to promote the collection of harmonized and well-measured phenotypes.[31]

The IGSS similarly stated:

> The purpose of this series of conferences is to bring together social and biological scientists with an interest in interdisciplinary work at the intersection of genetics and social science. Participants are typically working on problems that involve the use of genetic measures in social science datasets to enhance their ability to answer traditional social science questions. The guiding goals of the conference are to promote demographic and social science research and promote and expand the use of population data sources that include genetic information.[32]

These meetings drew more scientists working around the world into the fray. Take American-born sociologist Melinda Mills. Mills was working in the Netherlands at Groeningen University when she began to think that there were clear biological reasons for differential fertility rates among different nations. Drawing on a grant from the Dutch Science Foundation and support from the SSGAC network, she began shifting her focus toward genomics. In 2014, Mills was the first social genomicist to receive a grant from the European Union to build out social genomics (the European Research Council Consolidator Grant, a five-year grant of up to €2.75 million).[33] She has since formed SOCIOGENOME at Oxford University, where she has recruited Italian statistics expert Nicola Barban to oversee a social genomics program in reproductive health. In 2015,

SSGAC founder Koellinger received the same Consolidator Grant[34] to build out social genomics at his current home at Amsterdam University, and to establish a formal SSGAC headquarters. Koellinger has recruited Turkish econometrist Aysu Okbay to lead their educational genomics initiative. This rash of developments in the span of a decade elucidates the power and momentum of social genomics and its special brand of sociogenomic science.

Sociogenomic Science's Undisciplined Bent: A Structural Look at the Field

From this recent history, albeit brief, you get a sense of the who, what, and when of the science, but there are also the important questions of where and how, the "disciplinedness," so to speak. Today, social genomics resides in labs and departments around the world. Yet it still operates in the interstices of an array of social and biomedical sciences.

As we have seen, the first cluster of social genomics researchers emerged at Duke and UNC in the wake of Caspi and Moffitt's successes with the Dunedin cohort and the limited DNA sample release at Add Health. The tight network of interdisciplinary and interinstitutional partnerships between and within Duke and UNC, and their prescient sociogenomic agenda (best seen in the genomic redirect of the Carolina Population Center and the establishment of the sociogenomic Institute for Genome Sciences and Policy at Duke that followed immediately after the human genome was mapped), also fed into the establishment of these first hubs. Both UNC and Duke would foster genetics and social science working groups, which Add Health and these sociogenomic institutions would support and aid in developing. These arenas provided shelter for social scientists just getting started, who were asking the questions that had been burning in them for some time. It is important to note the demographical focus of this social genomics starter in the Research Triangle of North Carolina. Initial research did not spring from biomedical conundrums like cognitive function or disease, but rather from social conundrums like criminality and deviant social behavior.

Colorado also began developing into a hub, as Boardman pulled together demographers, sociologists, and behavior geneticists to assist him in rethinking race, neighborhood effects, and stressful life events. Unlike the Research Triangle, Colorado was not a home to transdisciplinary sociogenomic centers. But like the Research Triangle, Colorado was already established as a go-to place for learning about the genomics of behavior. The Institute for Behavior Genetics' two-week summer seminar in behavior genetics has been the training ground for economists, political scientists, and sociologists further afield for many years, and only increased in importance once social genomics came on the scene. Following the IGSS's launch in 2010, collaborations across these social science departments have only strengthened, as have the biodemographic emphases of the projects taking place therein. However, research has panned out from sociological concerns to health science interests in fertility and mental stability.

Nebraska with its Political Physiology Lab and Wisconsin with its Longitudinal Study cohort also became a hotbed of social genomics innovation in the early 2000s. Though everyone at this point was still using Add Health, the Midwest became a place where researchers could delve into basic questions about individual-level variation. It was in this configuration that biology shifted into focus. Researchers in Nebraska were starting with a panoply of brain questions aimed from a neuroscientific angle. And at Wisconsin, Freese and Fletcher were collaborating with evolutionary biologists and medical geneticists to rethink sociology and economics research in public health terms. Fletcher is one of several social genomics pioneers who have since gotten an appointment in the School of Medicine. These forays into biology have thus created lasting interdisciplinary ties between social and natural science departments at various campuses.

NYU became a critical part of the interdisciplinary "where" of social genomics when Conley made it the first campus to hire faculty specifically for their work on social genomics. As dean of social sciences, Conley supported the recruitment of political scientist Christopher Dawes as well as economist and SSGAC cofounder David Cesarini, while applying for an NSF IGERT grant[35] to formally establish a social genomics research institute. Cesarini and Dawes have led the field in its work with twins registries, most of which focus on biological and genetic contributors to disease, also bringing the field in closer relation to biomedical science.

With the foundation of the Gentrepreneur Consortium and later the SSGAC, Erasmus University also became a focal point in the interdisciplinary landscape of social genomics. As soon as it occurred to Thurik, Groenen, and Koellinger to work with genomics, each reached out to the biologists and medical geneticists at their institution. Their collaborators and the departments and institutes that they were housed in showed an eager interest in collaboration, and in 2007 that interest developed into the first full-fledged interdisciplinary research institute, the Erasmus University Rotterdam Institute for Behavior and Biology (EURIBEB). EURIBEB brought together the university's medical center, where its Dutch cohort was housed, with its school of economics. As its mission reads:

> EURIBEB investigates what makes people unique. We are an interdisciplinary research institute that links the social sciences with biology and health. Specifically, our research aims to integrate insights from biology (e.g. genetics, hormones, neurotransmitters, brain images), the social sciences (e.g. economics, psychology), and epidemiology (e.g. mental and cardio-vascular health) to address classic research questions in the "mother" disciplines, such as the causes of disease or the effects of public policy on school performance of children. The institute is a platform that inspires and nurtures collaborative research within this interdisciplinary context, bringing together leading scholars from the Netherlands and around the world.[36]

This institute has enabled the university's faculty to have protected time to focus their efforts on social genomics research, and has housed scholars from across the continent and beyond in their social genomics retraining efforts. The institute has served as the hub for genoeconomics in Europe, including but not limited to research done in Finland, Iceland, and Denmark. More importantly, it has provided a model for transdisciplinary innovation that does not privilege social or biomedical science, but rather nurtures the coproduction of the two into something new.

The transnationalization of social genomics has continued in the establishment of a British hub at Oxford and following Koellinger's move to Amsterdam. As noted, Mills put Oxford on the map when she established SOCIOGENOME with her ERC Consolidator Award.[37] Koellinger, who still has an affiliation with EURIBEB, also brought complexity to the landscape with his establishment of the new ERC SSGAC center.[38] These universities stand as two of Europe's most internationally populated academic institutions, and they house some of the most interdisciplinary centers for research in Europe. They also provide social genomics researchers cross-affiliation with world-leading medical and genomic research facilities. In 2015, Koellinger began working at the CTGLab (the Complex Genetics Traits lab housed within the Netherlands' top neuroscience and medical genetics lab).[39] The SSGAC and wider field affiliates have since trained their gaze on matters of cognitive function, schizophrenia, and Alzheimer's disease.

Thus the move to transnational and transdisciplinary science has also meant a move to medical aims and objectives.

Indeed, today medical genomics is recognizing the growing importance of social genomics and making room for it at the table. The first medical partnership with social genomics was built with personal genomics company 23andMe, so that members of the SSGAC could access its 23andWe research database.[40] This database is dedicated to crowdsourcing patient data on diseases like Parkinson's and inflammatory bowel disorder. SSGAC has also begun work with DeCode Genetics, the Iceland-based biotech company that was another founding member of the personal genomics industry.[41] Collaboration has proceeded similarly to that with twins registries and longitudinal cohort studies, where these companies have shared raw data so that SSGAC members can pool it with the rest. However, work with these companies has not just been business as usual. They have decidedly globalized the field further in its membership, reach, and operations. They have also furthered the field's turn to medical issues. A number of social genomics researchers have since been hired by world-leading medical labs. Just one example concerns Stanford School of Medicine's Ben Domingue. Domingue had been a part of Boardman's lab at Colorado and had been a central statistician in IGSS circles. Domingue has now joined social epidemiologists like David Rehnkopf in making their medical campus a new hub of social genomics.

Even more striking is the interest shown by arguably the world's top genomics lab in the world, the Broad Institute. The Broad was one of the main sequencing centers of the Human Genome Project and the global follow-up projects like the International HapMap Project and the 1000 Genomes Project.[42] Several of the Broad's founding directors were the leaders of these transnational projects, which have been mounted by hundreds of nations around the world. The Broad has also originated many of genomics' foundational technologies, methods used by genome scientists across the world. Therefore an invitation by the Broad means something huge for social genomics. As of 2015, social genomics researchers have collaborated in-house. In fact, the Broad is looking to them to develop the cutting edge of their behavioral research program.

There is only expansion, diversification, and transformation when it comes to social genomics—what the science does, what it stands for, and how it is done. Today, social genomics researchers work in social science departments, medical schools, demography centers, and on the premises of cohort studies. They work in low-tech offices just as much as in labs brimming with "high-throughput" medical equipment, in health institutes that collect a wide array of biomarker data, as well as in bench science facilities zoomed in on functional genomics. The "where" of social genomics is everywhere. It is no longer limited to social science or even academic spaces. Industry as well as government and supranational entities provide the resources and connective tissue for social genomicists to build out their habitats.

The fact that social genomics research is consortia driven, as opposed to discipline driven, makes for a field that is a-hierarchical. It is also driven by information and by information sharing. There is still no one scientist who is independently recruiting, genotyping, surveying, cataloging, and analyzing data. Social genomics researchers are either working with others (often globally) to do these things or they are only taking responsibility for the pooling and analyzing of the data. This means that researchers are free to travel and work. And they do. Meetings take place monthly, sometimes bimonthly, in different cities and countries across the world. Data analysis often piggybacks on conferences that scientists are presenting work at. This means that researchers just need secure Wi-Fi to operate. If I could draw a picture of the field's structure, it would look like a rhizome, a root system of a plant, as there are many crosscutting linkages between all points in the field.[43]

Rhizome Science: The Postgenomic Gold Standard

The story of social genomics is fascinating in its own right, but social genomics tells us something more. It is a window into the shifting structure of science at the dawn of the new millennium, in the postgenomic era. Analysts have heralded the fragmentation of scientific fields and scientific knowledge production in an age defined by globalization and information systems. But we have also witnessed the edification of interdisciplinary and transdisciplinary efforts, and the founding of new nontraditional science entities. Fields aren't simply disappearing. Rather, their nature is changing from something taut and solid to something amorphous, almost viscous.

It all began when governments began struggling to dominate the "knowledge economy" in the late twentieth century. Universities encouraged scientists to commodify their research. A "culture of commerce" washed over academic science at the same time as academic-industry boundaries dissolved.[44] Biotechnology was the poster child of next-millennium science.[45] In genomics, the scientists who were receiving the largest public grants, those who were working on national and international projects, formed their own companies. Since the turn of the millennium, these scientists and other scientific elites have been found heading up large academic labs, while sitting on the boards of large biotech firms and serving as CEOs and CSOs of their own start-ups.[46] While the days of scientists working largely removed from the world, holed up in their ivory labs, may not be completely behind us, it is no longer possible to assume that science works that way.

What has this meant for scientific fields? Fields are no longer necessarily located in academic departments that follow disciplinary strictures. They are no longer hermetically sealed off from other nonacademic professions, or from one another by analytical jurisdiction. Scientific fields emerge, grow, and evolve with respect to their interaction with industry, government, and other elements of the public and private sectors. And that's not even considering the ways that labs themselves have changed. Increasingly, scientific work is centered on digital technologies, software analysis, and big-data management. Who does this work and where it is done is wide open. It is as plausible that members of a field accomplish research remotely and with the assistance of coders and database administrators as in a lab enclosed by four tightly secured walls.

This informationalized, multiculti, network-based version of knowledge production isn't the science of the future. It's what today's leading organizations like the EU and the US Department of Health and Human Services are pushing for with their release of the NIH's "Interdisciplinary Research Consortium" or ERC's "Synergy Grant," an "interdisciplinary research environment" from the UK Medical General Council, and the National Academies' push for "facilitating interdisciplinary research.[47] These funding mechanisms ensure that the current generation of scientists will continue to seek interdisciplinary working conditions and relations, as well as train the future generations in transdisciplinary MOs.

Transdisciplinary science goals and objectives, and the ways social scientists in particular attempt to see them out, are also deeply shaped by the blatant divestment in social science that has struck the sciences in recent years. Government has responded to calls for cuts in research funding by making especially hard cuts in the social and behavioral sciences. The America COMPETES Reauthorization Act of 2015 has cut as much as 45 percent from the NSF budget for social, behavioral, and economic sciences.[48] Social scientists have had to go elsewhere for funding. However, it is not as easy to grab private funding as it once was. Many foundations that once solely funded social research have now turned to funding genetic research. One example regards the William T. Grant

Foundation, which in 2015 began funding Jason Fletcher's work on adolescent education and neuroplasticity as well as Joshua Brown's school environments project in 2014.[49] Another example concerns the Jacobs Foundation.[50] Jacobs has traditionally funded research on youth development but in 2010 awarded the Klaus J. Jacobs Research Prize to Moffitt and Caspi for their work on the MAOA gene.[51] Jacobs has since expanded its funding to epigenetic projects.[52] Funders have also launched educational opportunities like RAND's new Mini-Med School.[53] The Mini-Med School offers lectures on biology, patient care, genetics, psychiatry, and "how the practice of medicine can inform, and improve, social science research."[54]

As mentioned in the Introduction, I find it helpful to think about emerging scientific fields in sociological terms, from a "field analysis" or "field theory" perspective. Part of what this approach is about is reading the organization and structure of fields in terms of their identity and viability as a unique science. There may be a lot of infighting, but the emphasis is on "outfighting," and also consensus building and outreach, which goes into maintaining a science's position in the world. This approach is also about interpreting patterns in what scientists do in terms of the social networks that they work in and their field's collective struggles to remain legitimate and relevant.[55] In other words, it's about looking at scientists and their choices, whether career related, methodological, or specific to training and mentoring others, as about something bigger than them, something concerning the rewards and benefits that doing a particular science affords.

Indeed, at each meeting, I've witnessed conference organizers pose the question, "Are we at a point where what we are doing is creating a new discipline?" Organizers note the increasing visibility of the science, the renewed and newfound support from major national and international funders, and then move on to discuss where the science is in the matrix of sciences. Common topics to these opening sessions are how to run with the support social genomics is getting from biomedicine and how to, in the process, continue to demarcate what's special about social genomics. Finally, organizers sum up with the concepts and methodologies they see their science holding dominion over, despite "encroachment" from others.[56] This framing around uniqueness, demarcation, and viability shows that social genomics researchers are hard at work toward autonomizing into a field.

Still, I would argue that postgenomic field making isn't about drawing bright lines between domains of expertise as much as it's about capitalizing on broader reward structures in science, such as big pots of money offered by government health agencies and venture capital. Since science budgets have been slashed, researchers have had to survive by broadening research agendas, building bigger teams, bridging disciplinary gaps, and becoming creative in seeking funding. Solidifying a field is still a boon for getting taken seriously by the powers that be. But staking claim to a narrow agenda, as opposed to something that transcends disciplinary borders like method, is no longer the way autonomization is achieved.

It is an understatement to say in present-day science worldwide that having the support of the leading public science and health institutions is essential to the successful buildout of a field. As much as major agencies encourage science innovation under any guise, they put a high premium on bold attempts to integrate a number of priority areas that can advance healthcare. In the current climate, gene-environment science is the number one priority for most public health agencies around the world. It receives the lion's share of science funding in America, as well as other high-output governments like Japan, China, and the United Kingdom. And since the mid-2000s, when genomics moved into the postgenomic era of big data, an equally high premium has been put on statistical research and development and the integration of big-data bio-info, which goes beyond the standard numbers on an individual to capture their lifestyle, habits, and behavior.

So despite its decentered and fragmented arrangement, and its charge to be a science that handles anything social and genomic, social genomics is definitely a field that is coming into its own. Social genomics researchers define their collectivity as the preeminent science focused on the genetics of social phenomena. Take the SSGAC's collective "who we are" statement, which shows that the field demarcates itself from genetic and behavior sciences that do not focus on social science outcomes, and social sciences that do not study genetics, but has nothing more specific to its domain than method. It states:

> One major impetus for the formation of the SSGAC was the growing recognition that existing approaches to gene discovery in social science have often not produced replicable findings. The pilot project of the SSGAC on educational attainment has demonstrated the feasibility of an alternative, rigorous, large-data approach to social science genetics. Ongoing and future projects of the consortium will continue this approach to genetic discovery, and work on integrating the insights from these discoveries into medical and social scientific research.[57]

This elite group at the forefront of sociogenomic science sees its raison d'être as using the big-data social science know-how to improve biomedical knowledge. Unlike neuroscience or standard behavior genetics—fields that generally study small pedigrees of animals or cohorts of people with animal models, candidate genes, or neuroimaging approaches—social genomics researchers concern themselves with broad-scale demography and nationally representative polls. They thus fashion themselves as de novo "everything" science, a sort of social statistics meets bioinformatics dream team.

Social genomics is thus poised to reap the benefits of funding and honors that will be increasingly bestowed on transdisciplinary sciences that are able to prove their viability as lasting fields, chiefly because it is so ecumenical. With its mash-up of social science methods and genomic ones, the science is continuing to be pioneered by social scientists who have expertise over the very social phenomena that they study but are new to genomics, and by geneticists and doctoral and postdoctoral trainees of all backgrounds who assist them in the technicalities of what they do, making the field at its core not only interdisciplinary (partnering across disciplinary bounds) but transdisciplinary, or combinative of disciplinary approaches.

When we look at the raw demographics of the core scientists of this "field," "discipline," or "new science," as social genomics researchers but also their supporters and critics in the wider world call it, we see the transdisciplinary thrust all the clearer. At the heart of this growing field is a group of about one hundred economists, political scientists, and sociologists of all ages and career stages. One third of the field comprises economists. Another third consists of sociologists. The final third is split between political scientists and people with policy-related doctorates, such as public policy and criminology. Looking at scientists' BA fields, we see a similar breakdown for economics and sociology, but less representation for political science. Economists have the strongest correlation between BA and PhD field, with sociology coming in close behind, suggesting that the economists and sociologists comprised in the field are more likely to stay the disciplinary course in school than their political science peers. This continuity lends itself to recruitment of trainees, since a career track is established earlier on in these fields (one possible explanation for the smaller representation of the discipline of political science in the faculty that compose the field).

But what's important is the disciplinary distribution when it comes to recruitment and partnership. Single studies typically involve senior scientists from the array of contributing disciplines as well as junior scientists and research assistants who come from any background. Research teams bridge departments within universities, but more typically across them. So though traditional aspects of autonomization are underway, the field is nontraditionally solidifying as an interdiscipline with no limit to its analytical jurisdiction (what it's warranted to study). And rewards and benefits are acquired by way of transdisciplinary work around a capacious set of problems.

Despite these novel characteristics in postgenomic autonomization, there are still certain patterns that hold. For one, the field is evenly split among junior, midcareer, and senior scholars—a status important to the conditions for autonomization. In fact, one-third of all social genomics researchers are trainees at the advanced doctoral or postdoctoral stage. A second third are tenure-track assistant or associate-level professors. The final third are full professors who have considerable clout in their fields. Many of those senior scientists are gatekeepers, serving as editors of important journals in their home disciplines or departmental chairs and college deans of their universities. They are thus poised to recruit trainees and mobile junior faculty into permanent social genomics positions and to ensure the consideration of their work for publication. But what's really critical to take away from this career-stage breakdown is the fact that its evenness ensures that there are scientists at all levels to initiate and be initiated, to mentor and fund while expanding the jurisdiction of the science with new expertise, and to move up to make room for new recruits and the legacy of the science to continue on.

The nascent field also has an autonomization advantage in its breakdown by gender. Less than a quarter of the field comprises women.[58] Given that males suffer less of the structural and ideological sexism that plague women's careers, which weaken the continuity of a scientist's participation, male overrepresentation can enable, if not quicken, the field's formation. As studies on the sustained (and often worsening) gender gap in STEM fields have confirmed, the reason for the glaring gap in women's participation in the scientific workforce is that in the early to midcareer years, when scientists are building careers, women have disproportionate child-rearing and childbearing responsibilities. Women with children are also paid up to 23 percent less than men in the same positions—what social scientists often term "the mommy penalty." Thus even though women are overrepresented in most sciences in terms of holding PhDs, they remain underrepresented and unsustainably paid in employment. They are also less likely to be promoted fairly. And we know that an autonomizing field needs all hands on deck, and may not be able to withstand the exodus that results from institutionalized sexism of this nature.

Indeed, my survey of the field shows a successful rollout is taking place wherein male senior scientists create channels of knowledge and power transfers between themselves and new recruits. They start out by collaborating with midcareer scientists to forge new sociogenomic studies on phenomena for which they are already experts. Once their work is received well and published, and once they have established the viability of the research trajectory for themselves, they together begin recruiting junior folks. Eventually, a chain of expertise forms replete with the same kind of hierarchies and initiation strategies of other sciences. In my analysis of senior scientists, only a handful were women, suggesting that there have been very few female scientists who have been able to take that initial risk in bringing genomics to bear on their social science work. As a result, they have not been there to recruit or mentor more women. And while the midcareer male scientists who are now leading the field are deeply interested in shifting the balance, and are out there seeking female researchers, changes have not yet been reflected in the numbers, which may not change significantly until the field is stable and autonomous.

The same can be said for the field's racial breakdown. Even more striking than its gender gap is its race gap. Almost 90 percent of the field self-identifies as white. The remaining 10 percent self-identify as Asian, most hailing from and working in China. As I write this, only one social genomics researcher identifies as black. This researcher is junior, and has had moments in which he has considered leaving the field. Given that blacks represent a mere 6 percent of the STEM workforce, while making up 11 percent of the overall workforce, the existence of a race gap in a new field is not so surprising. But again, in the social sciences the disparity is not so great, suggesting that structural issues are at play. With respect to the field's autonomization and processes like recruitment, mentorship, and the availability of midcareer opportunities, I believe this racial homogeneity, as with gender, benefits the nascent field. Black, Latino, and members of the native populations of the Americas suffer from structural and ideological forms of discrimination that make it difficult for these scientists to sustain a career, especially during the early and midcareer stages, which autonomizing fields resolutely depend on.

With its members similarly positioned in these social identity terms, but diversely positioned in terms of academic position and expertise, social genomics is flexible yet stable. And it is exceedingly democratic. In conferences, social genomicists critique specifics of study design and methodology, but they do so in order to build consensus around social genomics concepts and methods. Even as debates heat up, participants give clear signals that their overarching goal is to support each other and raise a new generation of social genomics researchers. Critiques of controversial moves taken by scientists further afield are offered in light of ways to draw those scientists and their efforts into the fold. Critique of new scholars is exceptionally measured, and is often tied to instruction on alternative pathways to obtaining the scholars' desired results. Senior and midcareer scholars who are well versed in genomics nevertheless attend the genomics trainings that are provided at conferences in order to bring cohesion to the field's genomic approaches.

In sum, this sociogenomic field has a "flat," rhizomic structure based on interlocking lateral ties and a smoothly runged career ladder. It is in fact an autonomizing field, and not just some flash-in-the-pan intellectual movement or some passing research trend, but that is perhaps because autonomization is a different beast in this day and age. Social genomics researchers are invested in the cultivation of a lasting alternative to the current sciences in existence, but they do so as the methodological pioneers and analytical patron saints for a range of thriving disciplines. They hold dominion over nothing and yet everything at the same time.

Postgenomic field making and discipline building is, in effect, something new. Fields form from a firm basis of fluidity, while disciplines gain notoriety by virtue of transdisciplinarity. Yet there are traditional aspects of autonomization, like these demographic patterns, that remain strong. And there continue to be disciplinary walls, tendencies, interests, and hierarchies between the natural and social sciences that remain despite transdisciplinary efforts. That means we aren't yet able to say that we've completely transcended the "two cultures" divide between the natural and social sciences, wherein members of each tend to know little about the other, nor have we obliterated their competing interests. But we can, in the partnerships being made, see a move toward bridging that divide.

Molecularized Science Bestowed and Befitted from All Around

In later chapters I will present in more detail the reward structures of this interdiscipline, but it is important to give you a glimpse of the ways that postgenomic sciences with their nonspecific jurisdictions gain ground. Social genomics may show us how researchers are finding ways to get around the reward structures of their home disciplines, often traveling upstream to major health and funding agencies. But it is not just the pragmatism of the field's pioneers or their personal investment in mainstream pathways to scientific success in the postgenomic era that is the sole reason for the field's rapid success. The science has been propelled, and in many ways initiated, by the leading science institutions of our time.

When I spoke with the major funders in the United States—the NIH and NSF—I was surprised to learn that program officers actively sought out key social scientists who they thought could help with solving the big-data problems they faced. They knew that social scientists were the gatekeepers and key holders of many of the largest cohort studies in the world and were thus equipped to tackle the data issue. They had been using large datasets and individual-level information for decades. Plucking social scientists from their perch was a no-brainer for them.

But funders didn't just bring social scientists in to consult on specific methods and then return them to whence they came; they sponsored social scientists to retrain themselves in biomedical approaches. Social scientists were empowered to take the reins of their own biomedical studies. They were asked to bring their social science perspectives, especially their facility with theoretical thinking, to bear on major genomic initiatives.

As one director of a NIH office told me, he initially wanted to bridge statistical causality with biology causality models. He felt that "the huge opportunity of tomorrow" was a world in which "statistics and biology would become one."[59] This former sociologist had approached most of the field's pioneers, not waiting for them to come to him. He funded them so as to transform his home discipline and all the social science disciplines from the inside out.

Another director at the NSF told me that her office had been doing everything in its power to foster interdisciplinary funding for social science-led big-data research.[60] The resounding claim I heard from her and other funders was that they were in the business of pushing "high-risk high-reward" science. "High-risk high-reward" refers to research that funders are aware may not pan out; in other words, research that isn't just staking out a new arena of science, but rather conceives a new way of doing science through the radicalization of operations—an inquiry that has a high price tag but promises such a transformation toward efficiency and impacts that it is worth the gamble. "High-risk high-reward" continues to be the public agencies mantra.

Two major issues that funders have sought to get around have to do with the structural limitations of biomedical science. All genomic studies need to be replicated, but there are few incentives to replicate someone else's study. Original research articles that report a successful novel study are the currency of the biosciences, not false positives, errors, and nonreplications. There are also few incentives to share data in the biosciences. Thus it's really hard for labs to even begin to set up replication studies. Because many researchers are in soft-money positions, for which they must constantly get grants, it is in every best interest of scientists to just push forward with new studies. Funders see the more open-access, public data-sharing model of social scientists, who of course are usually in secure hard-money positions with tenure, as an opportunity to revolutionize science.

Meanwhile, the market has been nudging social genomics in its own way. In Chapter 8, I will present the litany of tests that have cropped up in the last couple decades. But it is important to pause here to consider the ways the overall growth in this sector feeds into the public and public health fervor around behavior genomics. Despite ups and downs in the biotech market, companies have steadily released personal genomics tests and have steadily increased their focus on behavior since the field first began selling genomewide genotyping services to the public. Learning about who you "really" are through mapping your DNA is now exceeding a twenty billion dollar industry.[61]

This brings me to the "why" of social genomics, and the "why now" of the broader paradigm of sociogenomics that I alluded to in the preceding chapter's concluding remarks. Both the field and the phenomenon arise now because of this crisis in the sciences and public health, a crisis that is taking place just as genomics is becoming a leader in public health in countries across the world. Transdisciplinary molecular science has become the focus of governments as well as corporate interests. Thus, as much as we can trace the building of the field and paradigm through individual actions, including career moves and scholarly aspirations, it is really these large-scale shifts that are driving the growth of social genomics and the expansion of sociogenomics. Molecularized gene-environment science, with its open doors and broad human interests, is the paradigm science of our day.

In our attempt to envision the stakes of all this, I want to leave you with an experience I had not so long ago when I walked into a sign-up tent for a marathon that was taking place in San Francisco. "Tent" is actually an understatement; I was walking into a hangar filled from one end to the other with runners, runner services, and marketing booths. The din was louder than a middle school yard at recess. There were booths for runners' foods and fitness regimes, hydration mechanisms and techy clothing. As I stumbled through aisle after aisle of schwag, I came upon a group of people sitting in chairs donating blood.

Just then, a woman approached me and asked if I wanted to donate to the cause. What cause was this? It was a sociogenomic fitness enterprise. By donating to their biobank, I would get the opportunity to purchase a spate of sociogenomic tests. Was I prone to accidents? Would I be good at certain kinds of sports? All I had to do was give blood and I could learn all about my own innate behavioral tendencies.

I asked the rep whether her organization was a nonprofit or affiliated with some academic research. She replied that it was neither. It was a proto-company, no investors yet, just an interdisciplinary team of scientists looking to establish proof of concept. This was a formless entity, sort of a virtual lab with no oversight. She assured me that the important thing to know was that this team had scientists from a range of backgrounds working on building the most holistic tool for self-knowledge and empowerment.

I politely declined to "donate" and stepped aside as a young couple rushed forward to give their blood to "the cause." As I walked away, I reflected on just how openness and flexibility can serve a science. Without a specific mandate, a specific jurisdictional aim, sociogenomic science has an unlimited power to make new truths about everything that we are as humans. But that only brought me to wonder: whose scientific standards would apply?[62] Whose ethics? Which of these entities involved in promoting the sociogenomic paradigm would stop to heed its social implications, and especially its nature-first leanings?

CHAPTER 3

TOWARD THE "DEEPER DESCRIPTIONS"

I'm a big believer that we should go to these deeper descriptions and
make connections with genes.

—Economist, European University

Whether in informal conversation or scientific publication, when social genomics researchers talk about what their science is about, they talk about the
search for a deeper meaning. They see social phenomena, or what they refer
to as "phenotypes," as but a superficial expression of the underlying processes
within us. In seeking to apprehend this deeper meaning, scientists have had to
create new concepts and methodological approaches to study design, data collection, and analysis.

This innovative search for deeper meaning has implications for the questions posed earlier, such as whether pairing genomics and social science—
creating something sociogenomic—changes the ways in which science backs
uncritically deterministic ideologies and the nature-versus-nurture binary that
has continued to reign. Is there something about the study of genes, the very
materiality of working with DNA and devising questions about it, that makes
scientists more data driven? Because as we saw with sciences like sociobiology and evolutionary psychology, much of that science has centered on theory
and conjecture, and not data. Scientists in those fields have not produced a list
of biological culprits for the behaviors they talk about, and they have left out
the body of environmental data pertinent to their inquiry. Is social genomics
different?

And if data are in fact driving the new sociogenomics, and fields like social
genomics are true gene-environment science, then we also must ask, how does
a gene-environment design change how research is done and what it means

for science and society? Does the social science leadership of social genomics make for a more socially and environmentally sensitive genomic science—a more balanced gene-environment science where disciplinary approaches thrive equitably? Or does the science stick to the sexier and more fundable side of things: genetic causes?

Here I elaborate the field's mainstay concepts and practices as seen in its seminal studies and as understood by the field's leadership. Examining study leaders' characterizations of their science in conjunction with the published record will show that the stakes of making social genomics look like a bona fide health science, and aligning with mainstream genomic efforts, are so high that this sociogenomic science ends up reproducing genetic determinism and nature-versus-nurture concepts. While researchers do not intend to lift the focus off of the environment, they are forced to recast social phenomena as "evolutionary phenotypes" so that they can make scientific claims that appear relevant to the vast array of biomedical and health sciences. They end up focusing on finding gene-gene interactions more than gene-environment ones as they brand their unique genetic association methodology. In this sense, the science stands to reproduce reductionisms that, while potentially profitable, threaten the true radicalness of this field's innovation.

Choosing a Good Phenotype

Let's begin by looking at the social phenomena that social genomics studies. Where do researchers get the phenomena that they choose to study, and how do they conceive them? Most derive their study phenomena from their prior social science work. Take an economist who once studied stock trading from a fiscal perspective, say by showing there to be income factors driving it, or a sociologist who has published widely on neighborhood effects on crime. These researchers still study these matters, yet from a newly minted gene-environment perspective.

In recasting social phenomena like stock trading and crime as the phenotypic end of the gene-environment relation (or as researchers put it the "genotype-phenotype" equation), researchers find ways to characterize those phenotypes as medically relevant. They then attempt to convince others in established medical fields to put their own imprimatur on these phenomena. As one American economist who leads the SSGAC exemplified in 2012, just as the field's momentous *Science* publication on educational attainment was in reviews,

> We started out studying the phenotypes that are convenient, that are available, that are of interest to social scientists—so educational attainment, subjective wellbeing, fertility. You know, in a lot of my other work, which is not genetics related, I study basic preferences like risk aversion and willingness to delay gratification, and trust, and fairness. And so those things I'm very interested in, but the data just aren't there yet to study those things. So we're trying to encourage medical datasets to collect that kind of data so we can study those variables.[1]

This researcher described crafting new definitions of phenotypes and presenting them to major epidemiological consortia so that they could be incorporated into multinational biomedical studies, an approach elaborated by a European colleague of his at the 2012 IGSS meeting, who explained that their need to stimulate "collaboration with [epidemiological] research groups around the world" was leading to robust characterizations of the biomedical meaning of each of these phenotypes.[2]

By the time the SSGAC's *Science* article hit the stands a year later, social genomics researchers were fully invested in what they saw to be a biomedical campaign.[3] As evident in reflections from one of their main collaborators in Europe, related to me at an SSGAC meeting in Rotterdam that year, researchers came to believe that converting social science phenomena into genetically associated phenotypes had become a pivotal approach in genetic epidemiology. They felt that social genomics was special because it possessed the power to unearth the genetic roots of social behavior and unleash the social potential of biomedicine. As one researcher told me, though he believes social genomics is just "basically satisfying human curiosity—we're just looking at this weird species called *Homo sapiens* and are trying to figure out who the hell we are and why we do the things that we do—"[4] social genomic phenotypes like educational attainment are truly "biologically proximate." By the article's June 2013 release, study leaders were arguing that educational attainment was "an obvious confounder for epidemiological studies" with a "causal influence on health—going to college actually buys you a couple years of additional lifetime."[5]

Indeed, that the organization's seminal publication appeared in *Science* established social genomics on the scene on par with leading genome projects, and suggested to the wider science community that graduating from high school and going to college (or not) were in fact serious health issues. It also solidified the genetics focus for the field, sending a message to others interested in social genomics that the wider science community was its primary audience. For example, in keeping with the focus of the journal, the paper cited literature that was biomedical and genomic as opposed to sociological or economic. And the SSGAC's press release cited "biological processes underlying learning, memory, reading disabilities and cognitive decline in the elderly" as the main benefits of the study.[6] Thus educational attainment was exemplarily rendered as a biological phenotype, no different from Alzheimer's or autism.

In translating social phenomena into phenotypes, researchers involved with the educational attainment study and beyond also recast phenomena in terms of broad evolutionary matters. In public conferences, private meetings, interviews, and publications, they repeated that the reason for studying the phenotypes they did was that they were interested in human drives that have been conserved over thousands of generations or, as one economist put it, "some underlying very fundamental dimension that influences a lot of different theories that we're concerned about as human beings."[7]

The essential and innate drives that concern social genomics researchers most are the drive for cooperation, for conflict, or the drive to take risks or avoid them. One scholar who works on political participation described to me a panoply of phenotypes related to voting behavior, characterizing them all in terms of altruism and empathy:

> I think there are certain types of people who just join groups or are active in causes in general. They're more willing to work with others or to become attached to a group if they maybe derive some social identity from that. You become active in a party or active in your church or active in this and active in that. And maybe you're the type of person, you have a certain type of personality, that kind of makes it easier for you to engage with other people. Or maybe you're just altruistic. But definitely engagement can be thought of that way. Like certain types of people are going to be more willing to engage and some type of people don't want to have anything to do with politics. And the degree to which they kind of do can be linked back to a personality trait or, you know, the way you feel about others, whether or not you're willing to help others, or you feel empathy towards others.[8]

A collaborator of his explained political behavior to me as being a special kind of essential human cooperative behavior, "and so you can learn a lot about political behavior by studying more broadly social behavior."[9] These scientists coded all forms of political engagement as expressions of an essential *biological* sociality that underlay being human.

In fact, researchers view social phenotypes in terms of a spectrum of innate social behavior. On one end of the spectrum are prosocial phenotypes, such as making friends and going to the polls. Researchers typically describe prosociality in the literature in this way:

> We hypothesize that genes may influence voting and political participation because they influence a generalized tendency to engage in prosocial behavior via their functional role in neurochemical processes. . . . These association studies give preliminary support to the conjecture that genes affect turnout and partisan attachment because they influence pro-social behaviors and attitudes that foster cooperation.[10]

On the other end of the spectrum lie antisocial phenotypes such as aggression, impulsivity, and violent behavior. As one political scientist related, antisocial phenotypes include "how people can commit violence either premeditated or not, either for what's believed to be normative good reasons or bad. And then those who impulsively do it and those who could get pleasure doing it."[11] Yet researchers read even benign actions or private feelings through the lens of antisociality, as exemplified in the previous article on voting: "Those who are overly sensitive to social conflict may choose to stay home and ignore politics, while less sensitive individuals will not take the potential emotional stress caused by the loss of their favorite candidates into consideration."[12] In this regard, seemingly neutral behaviors like shyness and declining to sign petitions become critical to the social genomic enterprise as signs of our species' evolution into a social animal.

Strikingly, social genomics researchers pride themselves on eliciting the deeper "conserved" (throughout generations of species development) meaning of our innate nature even though they do not attempt to directly measure evolution. As one review article on political behavior exemplifies, scientists instead use their imagination to come up with the evolutionary backstory to phenotypes: "From an evolutionary perspective, there would have been no selective pressure acting on specific political orientations or actions. However, it is possible to imagine that cooperative behaviors more generally did confer advantage on our ancestors to varying degrees."[13] Again, these researchers don't claim to research genes undergoing natural selection—those genes that have been passed down due to their ability to help individuals survive—by comparing the existence of specific genotypes in human populations across the globe. They don't use the biologist's gold standard tool of the evolutionary clock to date the arrival of said genes in the human species. Such methods have been successfully applied to phenotypes like being able to drink milk or breathe at exceptionally high altitudes, but they are not part of the social genomics toolkit.

As evidenced in one researcher's account of homophily, or the tendency for people to associate and bond with those similar to them, researchers see intuiting phenotypes in terms of evolutionary "models" and "working hypotheses" as the task at hand:

> So if you're a bacteria and you get a mutation and then all of a sudden you're reproducing, then all of the bacteria that are near you are probably going to have the same mutation, right? So some of it is just a consequence of location, but especially as you get to more complicated creatures, there is actually this process that goes beyond that, where we're deliberately seeking out others who are similar to us. And we were trying to figure out why this would be the case. And so we worked with [an esteemed Harvard evolutionary biologist] on this model of homophily, and it's a really simple model. The working hypothesis was that genes that regulate these neural systems are likely related to these political traits, as prosocial behaviors and attitudes underlie political behaviors such as turnout and partisan attachment.[14]

In this way, social genomics inserts itself at the nexus of evolutionary and social sciences, but with an emphasis on the innate human character. Scientists operate with an evolutionary imagination and intuition that permeate the field's every advance, making its science biomedically relevant even in the absence of a complete evidentiary basis. As such, the field's innovations around what social phenomena consist of and how they should be studied do not lead us away from genetic determinism, nature versus nurture, and nature first, but rather bring us into closer relation with them.

Determining the Risks in Our Genes

Take risk, for example. Nowhere is the tie to medicine, evolution, and a deeper essential sociality more evident than in the social genomic conceptualization of risk. Risk frames just about every phenomenon under study, gluing together a range of distant phenomena while rewriting them as inborn human traits. As one European economist demonstrated in his instruction on the genetics of entrepreneurship, the dominant interpretation of entrepreneurialism is that it comprises various forms of innate risk aversion and taking: "There is something called 'entrepreneurial orientation.' It consists of three parts: risk aversion, innovativeness, and pro-activeness. But these are somewhat deeper descriptions of human beings."[15] Another economist characterized temperament or personality as such: "We have temperaments measurements, if you know the [work at my home institute]. One way to measure people's personality is by asking questions related to risk and harm avoidance, persistence, extravagance, novelty seeking."[16] In both of these explanations, the risk concept sutures together an array of phenotypes to connote them with positive, innate prosocial qualities that portend good business habits and economic growth for society.

A working paper on genoeconomic risks (e.g., "genes [that may be] risk factors for poor educational outcomes, poor performance in the labor market, and consequently low levels of income") more broadly shows how risk glues together a suite of phenotypes, making them stand in for phenomena dominant in nonmedical and quasi-medical literatures such as economics and psychiatry:

> Various questionnaires ask about health-related decisions, such as smoking, drinking, eating habits, and conscientious health behaviors (e.g., getting regular check-ups). Each of these decisions reflects a tradeoff between the present and the future, and economic theory postulates that some individuals are more impulsive, or "impatient" in economics jargon.[17]

This paper, prepared for the National Academy of Sciences Workshop on Collecting and Utilizing Biological Indicators and Genetic Information in Social Science Surveys, names phenotypes as wide ranging as "labor supply and wealth accumulation . . . impulsiveness, risk aversion, and cognitive ability" as phenotypes "directly associated with underlying genetic propensities."[18]

Similarly, the article "Characterizing the Genetic Influences on Risk Aversion" states:

> Older Americans vary in their preparation for the financial burdens of retirement and old age . . . The increasing incorporation of genetic data into social science surveys has provided new opportunities to correlate genetic variants to observed social and economic traits. I exploit this opportunity to examine the genetic nature of risk preferences, preferences that are fundamental to most individual-level demographic events, including financial and labor market decisions, health behaviors, migration, and marriage and fertility.[19]

The study's frame reflects other suites of social phenomena that were related to me in interviews and observed in conference presentations and study meetings, phenomena that scientists band together under the banner of risk behaviors such as: (1) friendship, divorce, marriage, partnering, dating; (2) generosity, fairness preferences, trust, and financial risk taking; (3) religious affiliation, attendance, fundamentalism, observance of holidays, membership in youth or community church groups; and (4) time discounting (i.e., delaying gratification), income, occupation, and trust.[20]

Risk is thus what sociologists call a "mutable mobile"[21]—a successful and durable frame that thrives off of fluidity and an ability to play many roles. Sometimes "risk" is an umbrella term for all prosocial or antisocial tendencies. Other times it is a sister term alongside a subset of nesting preferences. As exemplified in the paper "The Genetic Architecture of Economic and Political Preferences," researchers use risk, preference, and behavior interchangeably:

> We focus on preferences because they are fundamental building blocks in the models that economists and political scientists use to predict behavior. For example, measures of risk preferences predict diverse risky behaviors, such as smoking, drinking, and holding stocks rather than bonds. Experimentally elicited patience predicts body mass index, smoking behavior, and exercise. Political preferences similarly predict a wide range of political behaviors, including voting and monetary campaign contributions, as well as campaign activities like volunteering, attending rallies, and displaying yard signs.[22]

From holding stocks to displaying yard signs, expressions of risk are understood as nesting indicators of a deeper essential human typology.

Addiction studies and other health-related analyses, such as obesity and smoking behavior, play an interesting role in social genomics, since many researchers have come to study the genomics of risk from these quasi-medical angles yet with a decidedly nonmedical, evolutionary bent. Again, here nature is the explanatory factor, as this genosociology article on adolescent substance use in school environments exemplifies:

> Although a large number of genes and social influences have been linked to tobacco and alcohol use, relatively little research has investigated multilevel gene-environment interactions for these outcomes, and none have taken such an approach to studying tobacco and alcohol co-use. However, outside of substance use studies, a large body of research has demonstrated the potential importance

of 5HTTLPR (a polymorphic region in the SLC6A4 gene) for a large number of outcomes such as psychopathy, impulsivity, alcoholism, and violence.[23]

The authors of this paper not only take the focus off the original phenomena of school environments and their influence over substance use in this search for deeper evolutionary meaning, but also move from any gene-environment analysis to gene-gene analysis (in this case, 5HTTLPR and SLC6A4). Like others across the field, their interest lies in the relationships between the genes they believe are responsible for the behaviors they see.

Similarly, in an interview in 2014 following an IGSS meeting, a researcher who focuses on cocaine addiction said that risky sexual behavior would "be the most natural extension of what I have done on those cocaine uses and, say, a number of factorial partners."[24] She alluded to some underlying font of risk that makes some people addicted to drugs as well as promiscuous. Researchers like her assume that they will find a number of genetic culprits, so they search for those markers, leaving the complexity of the environment aside.

Despite filtering everything through a gene-gene prism of innate risk in place of a gene-environment analysis that takes the social and built environment equally seriously, many researchers nevertheless draw our attention to social policy, even going so far as to make policy recommendations. One way they do so is by interpreting phenotypes in terms of their potential to create social disruption or deviance, and charging possessors of their related genotypes with the potential to harm others. For example, in a study on prenatal smoking, a team of scientists warn that pregnant smokers in possession of certain addiction genes will harm their babies, setting off a chain of future societal pathologies such as "developing behavioral problems later in childhood, participating in criminal behavior, and lifetime nicotine dependence [and] also increased risks of language problems, hyperactivity, fearfulness, and not getting along with peers."[25] In another study, they compare gene-gene interactions between variants associated with prenatal smoking, alcohol use, and obesity in order to characterize other harmful implications, like making kids overweight and antisocial.[26] The message they send to policymakers is to stop these genetically programmed pregnant women from smoking, abusing substances, and being overweight, or else all of society will suffer.

Another policy-focused researcher told me that smoking and obesity are really just addictive self-harm, and that self-harm that is transmitted over generations holds high social costs that our healthcare system can't afford:

I think of it more as a social problem, but it's the pathogenesis of the intergenerational transmission of disease and disadvantage. So we have social strains in this country that are resistant to our best efforts to change them. Some people are born sick and some people are born rich and healthy. So the question I'm trying to ask is, what can we do about that? How can we try and help people that are born into less advantaged circumstances to grow up and live healthy lives?[27]

In essence, this researcher sees the social genomicist's job as one of finding the genes that mediate individual behavioral responses to the social order, including policy and culture, and adapting policy and culture to those genetic differences.[28]

Even when researchers aren't detailing specific policy solutions, most imply that society should make genetics-based policy. In the article "Age at First Sexual Intercourse, Genes, and Social Context," researchers look for genes responsible for risky sexual behavior—"relative risks of first sex by age"—by fine-mapping the "alcoholism gene."[29] They warn that sexual behavior can be caused by feedback loops in the brain that signal it to produce more dopamine: "Such increased dopamine levels could result from a variety of compulsive, impulsive, and addictive behaviors, including novelty seeking, ADHD, risky sexual behavior, substance abuse, alcoholism, smoking, binge eating, and compulsive gambling."[30] In linking risky sexual behavior to this litany of other risky behaviors spurred by specific dopamine receptors, they suggest that possessors of the "risky genotype" will need to be treated according to these specifics of their biology, especially since an earlier date of first intercourse often leads to abandonment of contraception, the spread of sexually transmitted diseases, and an onslaught of teenage pregnancy. Finally, they imply that these social ills will not abate without knowledge of who possesses these "risky genotypes." They do not offer any environmental counsel, only a yet unfulfilled promise of future research on environmental factors.[31]

Capturing a Good Genotype

In various meetings and virtual hangouts, social genomics study leaders shared with me a number of novel phenotypes that they are just beginning to explore, things like life satisfaction, income, occupation, and partner choice. Musing on phenotype always inadvertently led to reflection on genotype, namely how to get at it. I learned that study designers use three main avenues: twin studies, candidate gene studies, and genomewide association, or "GWAS."[32] In their original uses of these methods, we see the same reversion to "nature first" that takes place with their conceptual work.

Proving biological heritability is the first step in any social genomic study, because unlike with health phenomena it's hard to justify studying the genetics of behaviors that appear to be due to the environment. Researchers most often refer to prior twin studies to establish heritability at 20 percent or more, as in this example taken from a report prepared for the National Institute of Justice titled "The Intersection of Genes, the Environment, and Crime and Delinquency: A Longitudinal Study of Offending":

> If conduct disorder is genetically influenced, then MZ [monozygotic or identical] twins, whose co-twin has been diagnosed with conduct disorder, will have the greatest genetic risk for also developing conduct disorder. DZ [dizygotic or fraternal] twins whose co-twin has been categorized as having conduct disorder will have a lower genetic risk for also being characterized as having conduct disorder. DZ twins whose co-twin does not have conduct disorder will have an even lower genetic risk score. And, finally, MZ twins whose co-twin has not been designated as having conduct disorder will have the lowest genetic risk for conduct disorder.[33]

However, social genomics researchers also conduct twin studies themselves. For instance, "Stressful Life Events and Depression among Adolescent Twin Pairs" represents the results of an original twin study to find genetic factors responsible for stress exposure. As its authors argue: "SLEs [stressful live events] are generally treated as exogenous shocks, stress is characterized as something that *happens to* people. However, it is also possible that stress exposure and poor mental health are both derived in part from the same unobserved source."[34] Because prior research reports heritability estimates for stressful life events on the order of 17–45 percent, and because "all self-reported environmental measures [parental warmth, family cohesion, family conflict, family organization, social integration, and the perception that friends have problems] evidence additive genetic variation," these researchers use survey and genotype data taken from the twins population in Waves I-III of Add Health to broaden analysis to a nationally representative sample.

Overall researchers prefer to conduct twins analysis in preparation for other genomic approaches. For example, when a team of genoeconomists used genotyped subjects from the Swedish Twin Registry who were surveyed with "Screening across the Lifespan Twin Younger" cohort (SALTY), "a rich set of questions measuring economic and political preferences" (e.g., risk aversion, patience, trust, fair-mindedness, and attitudes about immigration/crime, economic policy, environmentalism, and feminism/equality), they qualified their choice of conducting their own twins analysis by stating:

> For comparability with previous work and with our other estimates, we report twin-based estimates of heritability from this new sample, and we confirm moderate (30–40%) twin-based heritability estimates for these traits. However, our main focus is on using the dense single nucleotide polymorphism (SNP) ["genetic variant"] data to learn about the genetic architecture of these traits.[35]

Here and in private meetings, these researchers reminded others in the field that twins analysis doesn't actually get at the genetic causal factors—the very genes responsible for behaviors—it only establishes heritability.

Many study leaders have advanced a "candidate gene" approach as a way of finding genetic causal factors. Candidate gene studies fine-map specific genes that are believed to be responsible for the trait in question. As explained by one team of researchers:

> Twin studies have already established that genetic factors account for a significant proportion of the variation in antisocial behaviors, including substance abuse, impulsivity, criminality, precocious sexuality, and a combination of these behaviors called antisocial personality disorder (ASPD). However, twin studies cannot establish which genes are implicated. It is likely that dozens, if not hundreds of genes influence sociability. As a result, scientists typically start with "candidate" genes that are known to influence related behaviors or processes in the body. For social behavior, this means focusing on genes that affect brain development, neurotransmitter synthesis and reception, hormone regulation, and transcription factors.[36]

Up till now, candidate genes such as the two advanced in this paper, 5HTT and MAOA, and the addiction-related DRD2 and DRD4 alleles mentioned above, have been scientists' main targets. All of these genes have been shown to influence dopamine, serotonin, and other hormone metabolism, and to be associated with social behavior. As such, they are interpreted as the key to prosocial and antisocial genotypes.

For many pursuers of social genomics, however, a candidate gene approach also stops short of the gold standard of genomic science and medicine—the GWAS (again, "genomewide association") approach in which entire genomes are analyzed for causal factors. And these scientists desire their methods to reflect the genomic mainstream. As the study on risk aversion quoted above sums:

> Candidate gene studies provide complementary evidence to the estimates of overall heritability provided by twin studies. They examine associations between survey measures and forms of a small number of genes chosen on the basis of prior knowledge about biological pathways. In work with candidate genes, handfuls rather than millions of statistical significance tests are at stake, so much weaker associations can be found to pass thresholds for statistical significance.[37]

This study reports on genotypes of 10,455 adults from the US Health and Retirement Study. Colleagues at SSGAC produce studies that average on 150,000 subjects. Indeed, in SSGAC conferences and study meetings, researchers constantly worry about insufficient sample sizes, and they express dismay at the lack of rigor in candidate gene studies.[38]

Whether adopting a candidate gene or genomewide approach, social genomic innovations in gene-environment research are presently only continuing to tip the balance toward genetic explanation and away from deep environmental analysis. Studies map genes and intergene interactions, but they do not illuminate complex social-environmental factors or the ways in which those factors together create the social conditions for behavior—the original interests of most of these scientists. Yet in moving afield from their original interests and explanatory factors, social genomics researchers make their research interesting to a broader array of sciences and health agencies. They become relevant to behavior geneticists who work with animal models, as well as medical geneticists who work on human disease. They become conversant with neurology, oncology, and other scientific fields that characterize brain activity and cellular pathways. Meanwhile, they find favor with the social science establishment for bringing genomic methods into the domain of social science.[39] They align with a wider array of long-standing fields than they would if they were to present social science as the main attraction. In the current climate, where a tradeoff between transdisciplinarity and empirical complexity continues to reign, transdisciplinarity wins the day.

Building a Sociogenomic Methodology

The specifics of the social genomics methodology provide more evidence of the ways that social research aims get pressed into a gene-focused model of research. In trying to apply GWAS to study phenotypes that have been ascertained by survey research, study scientists must get genetic and survey data from a robust sample. While social surveys with less than one thousand respondents are often considered robust, GWAS requires thousands of study participants. Researchers look to national cohort studies, such as Fragile Families and Add Health, that collect substantial amounts of DNA and survey all their subjects for their main sources of data. But "getting the numbers right" often necessitates combining multiple waves of several cohorts' data. As one doctoral trainee in genoeconomics put it:

> The main thing it boils down to is sample size. So you accept all of those just to have a much bigger sample size, because the idea is that in the end the sample size compensates for all kinds of noise that you may encounter. So yeah, that's the idea. We get all kinds of acceptable measures just to increase sample size. That is the most important thing in the end.[40]

An economist who focuses on entrepreneurship lamented these limits, while sharing his dream that social genomics team leaders would themselves run cohort studies in the future:

> The sad thing is, on this level, we don't have large datasets yet. Maybe with the big biobanks, which are around the corner, maybe this will be a solution. But that'd be inherited from the fact that the datasets had been set up for medical reasons. As some sort of sideshow some of them have asked, "Oh, by the way, what did you do with your life? Did you graduate from high school? Yes or no? Or did you go to university? Yes or no?" In the future hopefully it's the social economic researchers which will ask the usual questions and then say, "Oh, by the way, can we have your blood?"[41]

Many emerging researchers like him explained ways that they push cohort studies to see DNA collection as the wave of the future. His remarks evince the eagerness researchers have for social genomic interests to lead biomedical data collection, but also the ways the medical or disease analysis would fade from view if this were to happen.[42] More importantly, they show how study scientists are primarily concerned with the genomic end of sampling. For the most part, social genomics researchers do not attempt to innovate around research population inclusion, convincing various cohort studies to let them pool and analyze data across multiple studies in order to survey and characterize phenotypes in new and interesting ways. Rather they ask cohort studies to do more for them so that they can do more with DNA.

This emphasis on genomic innovation, as opposed to social scientific innovation, is most evident in the way that study scientists use surveys. Researchers often contribute nothing new to the social side of their studies. Once cohort studies are selected for sampling, they simply use the survey measures already at play in them. For example, in the aforementioned study on time of first sexual intercourse, researchers deployed these Add Health survey measures:

> "Have you ever had sexual intercourse? When we say sexual intercourse, we mean when a male inserts his penis into a female's vagina." If the respondent's answer was yes, he or she was then asked, "In what month and year did you have sexual intercourse for the very first time?"[43]

In an SSGAC study on subjective well-being, researchers imported preexisting questions like: "Do you enjoy life? Are you happy?"[44] Likewise, the risk aversion study of financial investing used these measures for risk from the Health and Retirement Study:

> Suppose that you are the only income earner in the family, and have a good job guaranteed to give you your current (family) income every year for life. You are given the opportunity to take a new and equally good job, with a 50–50 chance it will double your (family) income and a 50–50 chance that it will cut your (family) income by a third. Would you take the new job?[45]

As researchers import extant survey measures as is, they create new metrics that can quantitatively translate social measures into genetic ones. The SSGAC study of educational attainment, for instance, was measured by coding "study-specific measures using the International Standard Classification of Education (1997) scale (14)"[46] and creating a continuous variable for years of schooling and a binary variable for completion of college that could be read as a simplified risk score.[47] Their study on subjective well-being also stratified responses into positive affect and feeling happy, satisfaction with life, and overall well-being, with each registering approximately 115,000, 84,000, and 150,000 responses respectively into this kind of score. Likewise, for a study on MAOA and delinquent behavior, researchers divided Add Health survey measures into nonviolent and violent types drawing lines between stealing or drug dealing and stabbing or "deliberately damaging property."[48] All of these scores allow researchers to flatten environmental diversity in order to feed it through GWAS software as simple phenotypic data, the way that body mass index or height gets fed into GWAS equations as binary code for over- or underweight, tall or short.

Even when researchers add new social study components, they add them as is and then feed scores into genomic analysis. A common component to include in a social genomic study is an experimental economic or political game, such as the "dictator game" or the "ultimatum game," or as one economist called it, "the usual game-playing questions of how much would people be ready to pay for being able to participate in a game where you can win or lose from different amounts." In the *Science* publication that SSGAC members refer to as "Educational Attainment 1.0," the consortium forecasts the following:

> We also plan to add standard experimental measures of impulsive and risk-averse preferences to the next wave of the AGES-Reykjavik study. These protocols ask participants to choose between immediate vs. delayed monetary rewards or to choose between certain vs. risky monetary rewards. These choices are played out with real monetary stakes.[49]

As others expressed, these study leaders do not purport to revolutionize methods for getting at impulsivity and risk aversion, but rather promise to add standard, "tried and true" measures.[50]

There is one additional method that several social genomics researchers use as an alternative to twin studies, candidate gene studies, and GWAS that also demonstrates a digression from social science empirics and innovation: telomere analysis. Telomeres are the sequences of DNA that cap the end of chromosomes and shorten with each cellular replication. Genomic research has found several genetic markers that influence an individual's telomere length. As such, they are simultaneously an indicator of biological aging and genetic determinants of physiological deterioration. Social genomics studies use the standard telomere gene-mapping approach to characterize social weathering or resilience to stressful life events. A telling example concerns research on telomere erosion and military service. In a 2013 study, one team of researchers found that for all postdeployment soldiers each month of military service equated to one year of nonmilitary life lived, as measured by declines in telomere length. But soldiers who reported having post-traumatic stress disorder (PTSD) and those who had attempted suicide since deployment had double the rate of decline, suggesting a stronger genetic predisposition to vulnerability.[51] Telomere analysis continues to be debated in social genomics for its use of sample sizes in the hundreds as opposed to thousands.[52] However, research into occupation, fertility, and gender preferences is underway across all areas of the field.[53]

Taken together, these methodological innovations all point to the same drive forward toward not only a nature-first picture but also a more genes-first view of human existence, one that elides deep analysis of environment or context in favor of DNA, as is demanded by the scientific mainstream.

Path Dependency, Pleiotropy, and Other "Genes First" Precepts

No matter how gene-focused social genomic methodology is, combining cohort study surveys and genomic analyses transforms the meaning of genetic and social science and ushers in new methodological constructs that shift the terms of these sciences. As cohort studies are rationalized with genomic protocols, they become an arm of genomics. But genomics, now approached in terms of waves of self-reported data, also becomes rationalized with the logic of social science cohort studies as well. This is the boon of sociogenomic science, a transdisciplinary science in which new methodological constructs enable researchers to tackle phenotypes as continuous, fuzzy-set variables so that they can claim some degree of predictive power over social outcomes. Yet though they are open-ended in a way more akin to typical social science measures rather than genomic ones, these novel constructs also reinforce the genes-first assumptions evident in many of the field's foundational concepts and methods.

A path-dependent, or "life course," perspective taken from the social sciences is one of the central innovations of the social genomic methodology. Researchers approach phenotypes as malleable over the life course, most relevant to specific parts of the life course, and influential over future behavior. So as one sociologist who researches "anything that might be impacted by parents" (e.g., "financial outcomes, education, health") maintains, measuring subject behavior requires a path-dependent, age-conscious approach.[54] Another researcher active in health policy thinks of genetic risk as path dependent and developmental throughout life:

> I think the broad message is that the genetic risks that we are discovering primarily in cohorts of middle-age adults are manifested much earlier in the life course, and those early manifestations appear to be critical in determining

whether the genetic risk is going to contribute to disease in adulthood or not, with the suggestion that if we could intervene early to prevent or mitigate these early manifestations, we might in fact be able to sort of treat the genetic risk in a lasting way across the lifespan. So DNA isn't destiny; it is our malleable risk exposure. We can do something about this as a public health and public policy apparatus.[55]

This researcher believes that the life-course approach is what makes social genomics unique from other fields that study similar problems: "That's not really what people in genetics doing obesity are about. They are interested in understanding the genetics of obesity. I'm interested in something a little bit different from that."[56] Counter to the dominant approach in genomics, these researchers capitalize on cohort study structures to home in on subject populations that can represent certain parts of the life course, such as adolescence or early childhood, and they tailor their study component additions to a path-dependent framework that can encompass stressful life events, abuse, parental use of health services, and other environmental dimensions that may bear on behavior. In fact, at one scientific conference, attendees discussed the possibility of creating a behavioral life-course model that would explain health-risk path dependency—"an 'addiction' or 'health stock.'"[57]

Yet the focus on life history and specificity of life stages still leads scientists further into gene-gene analysis. Study leaders interpret the task of studying multiple phenotypes as a charge to find new ways of measuring pleiotropy, the process by which one gene affects multiple phenotypes. One article on attention deficit and hyperactive disorder (ADHD) and educational attainment describes how researchers see this conundrum:

> When there is pleiotropy, some of the genetic variants have a true direct biological influence on both phenotypes. For example, a number of molecular genetic studies demonstrated that there are some genetic variants on chromosome 6, 13 and 14 that have an effect on both reading disability and ADHD. The pleiotropic effect of a genetic variant can occur when a gene is involved in multiple biological pathways or the same biological pathway has different effects on the associated phenotypes. For example, dysfunction in the dopaminergic pathway has been implicated in the development of ADHD and this pathway has also been associated with cognitive function. Alternatively, it may be that the genetic association appears because there are genetic variants influencing ADHD and being genetically predisposed to ADHD makes it harder to concentrate at school,

leading to lower educational achievement. Or, the other way around, children who have problems keeping up in school display, perhaps out of boredom and frustration, ADHD symptoms.[58]

Pleiotropy is assumed whenever a GWAS computer analysis brings up a "hit" on a genetic variant, whether or not the study has proved there to be a direct relationship between the phenotypes.

Researchers use a biostatistical software program (what's known as bivariate GREML analysis, or Genome-wide complex trait analysis, "GCTA") to measure the relationship of phenotypes under the assumed pleiotropy. As a graduate student explained:

> So what people in quantitative genetics and genetics in general used to do when they wanted to infer the proportion or variation in a trait which could be attributed to genetic variation, so-called heritability, what they used to do was use twins data in order to infer this. So monozygotic twins are more similar than dizygotic twins and that's a basis for a heritability estimate. But in GCTA, you basically replace it with raw genetic data and infer it from that. So what you basically try to do there is see, does genetic similarity lead to phenotypic similarity? . . . Are the genes that contribute to height the same genes that contribute to BMI [body mass index]? Similarly, are the genes, so to speak, that contribute to educational attainment similar to the genes that contribute to self-employment or risk preferences or anything else?[59]

These remarks show that social genomics researchers allow the genomic software to tell them about the relationship between phenotypes. From there, they require no further social-environmental analysis. As the student further explained, they even allow the software to determine how phenotypes play out along ethnic lines:

> Another thing when we come to the gene-environment is actually to say, "Are the genes that contribute to educational attainment in Rotterdam the same genes that contribute to educational attainment in the US or in Sweden?" So then you can also say if this phenotype, educational attainment in Rotterdam, is something different from educational attainment in Sweden, is something different from educational attainment in the US. And then you actually try to test the assumption that there is genetic correlation.[60]

The comparisons this researcher attests to are an about-face from standard social scientific ways of imputing whether social phenomena affect ethnic groups in different ways, measures that take socioeconomic status, neighborhoods, and social policies into account.

Indeed, the open-ended fuzziness that characterizes the social genomic modus operandi seems to always refer back to some yet-to-be-determined genetic source, as evident in the field's foundational concept: the endophenotype. The endophenotype is an unmeasured trait that researchers believe to be behind a study's measured phenotypes. Researchers work with the assumption that there will always be a number of "unobservables" in every study.[61] In fact, the large majority of researchers in this field maintain that each and every social genomic study produces findings on an endophenotype. As one economist described:

> It starts with the genes and in the end there is some choice or behavior or decision, like entrepreneurship, or go to the university, or get married, or whatever. But in this pathway, just before the last step to something everybody can see and measure, there is the world of psychology.[62]

A political scientist put it this way:

> Ultimately, you can measure people's participation or you could ask them what they do. Maybe on the other end, it's genes, but why? What comes in between? It could be that for some people, they think of being active as a way to help society or at least help people within their group. So a Democrat wants to help a fellow Democrat, or like an older person wants to help a fellow older person. So really for me, it's like trying to fill in those blanks by collecting more information, . . . like taking a step back and saying, "What's something that's a little bit more basic? And that ultimately influences the phenotype?"[63]

This researcher suggested innate personality traits and cognitive ability as some potential culprits, but all studies characterize endophenotypes thusly—as intrinsic biological traits upstream of the social phenotype at play.

The other major fuzzy-set methodological construct that social genomics researchers deploy—one which is entirely gene focused—is the polygenic risk score, also called "genetic risk score" or just "risk score." When the genetic variants they find display miniscule effect sizes, as social genomic analyses regularly do, researchers create a composite score of the cumulative effects of the identified genetic variants. As one sociologist explained:

> Genes do way better when you take them in aggregate together. And they do way better than our sociological variables toward predicting the social behaviors. We

can predict education from someone's genome. Let me just back up: maybe not education but with a lot of things; way better from the genetic information than you could from any number of survey questions that you add or added all together. Our score is just higher to explain variations.[64]

In essence, a polygenic risk score allows researchers to assign a "genetic burden" to a phenotype for which they would otherwise never have sufficient numbers.[65] This gives researchers the confidence to draw conclusions about subjects' future choices, actions, and behaviors and to theorize about policy. As this sociologist continued:

> So imagine in [a GWAS of menopause] like we predicted, we've got something that was like 75 percent r-squared. Pretty damn predictive of age of menopause! I think an eighteen- to twenty-year-old woman would want to know, with 75 percent certainty what her predicted age of menopause is or her end of her fertility, so that she can make social and economic decisions based on that. Like maybe she'll think, "Well, I'm predicted to have a short fertility span so I'm going to have babies now, you know, while I'm still in school."[66]

Though polygenic risk scores normally only predict 2–3 percent of the variance of a trait, not 75 percent as this researcher purports, the open-endedness of the construct empowers social genomics scientists to think big and talk big about the potential impact of the causal variables they are identifying.

Polygenic risk scores also give researchers the confidence to draw conclusions about the endophenotypes just mentioned (those unmeasured biological causal factors assumed to be influencing measured phenotypes), and to justify their theories and methods. For instance, in the SSGAC's follow-up study on educational attainment (what they call "Educational Attainment 2.0"), researchers examined educational attainment alongside a phenotype for which a great deal of genomic research already exists: cognitive function. Comparison of polygenetic risk scores showed that the genetic variants found in their GWAS explained 2 percent of the variance in educational attainment, but 3 percent of the variance in cognitive function, leading researchers to draw the conclusion that cognitive ability was a mediating trait responsible for both phenotypes. And as one SSGAC leader described regarding educational attainment:

> You can basically rule out an environmental confounder if you can show that within families, if you just look, for example, at full sibs—if you can show that within a pair of full sibs the genetic variance still has the same effect. Although these people

have been living in exactly the same family environment, you can basically be sure that the effect is really genetic and not driven by some unobserved environmental confounder, right? So the thing is, this is exactly what we wanted to do, but the effect sizes we found were so tiny that there is just not a single sample in the world where you have enough full sibs or family members that will allow us to do that. So instead of looking at the individual genetic variance, we looked at the polygenic scores—the linear combination of all the genetic variance together. And if you do that, then you do have power to do that. So we did that in two family samples—the Queensland Institute of Medical Research, they have a twin sample, and the Swedish Twin Registry as well—and basically we tried to predict differences in educational attainment within pairs of brothers or pairs of sisters. We were basically trying to predict differences in educational attainment within these pairs using our polygenic score, and it worked![67]

Here the polygenic risk score is a useful fact-checker for a study that assists with the unique challenges social genomics studies face due to their sample size needs.

Finally, polygenic risk scores allow social genomics researchers to explain missing heritability, failures for GWAS to turn up genetic determinants, and failures in replication. Take for example these summary remarks from an inconclusive study on entrepreneurship:

We report results from the first large-scale collaboration that studies the molecular genetic architecture of an economic variable—entrepreneurship—that was operationalized using self-employment, a widely available proxy. Our results suggest that common SNPs [single nucleotide polymorphisms, or genetic variants] when considered jointly explain about half of the narrow-sense heritability of self-employment estimated in twin data ($\sigma_g2/\sigma_p2 = 25\%$, $h^2 = 55\%$). However, a meta-analysis of genome-wide association studies across sixteen studies comprising 50,627 participants did not identify genome-wide significant SNPs. 58 SNPs with $p<10^{-5}$ were tested in a replication sample (n = 3,271), but none replicated. Furthermore, a gene-based test shows that none of the genes that were previously suggested in the literature to influence entrepreneurship reveal significant associations. Finally, SNP-based genetic scores that use results from the meta-analysis capture less than 0.2% of the variance in self-employment in an independent sample ($p<0.039$). Our results are consistent with a highly polygenic molecular genetic architecture of self-employment, with many genetic variants of small effect. Although self-employment is a multi-faceted, heavily environmentally influenced, and biologically distal trait, our results are similar to those for other genetically complex and biologically more proximate outcomes, such as height, intelligence, personality, and several diseases.[68]

In other words, a polygenic risk score of nearly 0 percent is justification for further analysis of the genetic determinants of traits. This optimism is reflected in the IGSS's 2014 special issue in *Biodemography*, in which scientists and the issue's editors proffer polygenic risk scores of 0–3 percent as springboards for further analysis,[69] and is encapsulated in the concluding remarks of the SSGAC's 2014 *Psychological Science* address to the discipline of psychology, where scientists state:

> The polygenic score explored here has modest explanatory power ($R^2 \approx 2\%$), but when the weights for constructing the score are estimated in larger samples, the explanatory power will be much greater. For example, Rietveld et al. estimated that a polygenic score constructed using results from a discovery sample of 500,000 individuals will have an R^2 of approximately 12%. We anticipate that such sample sizes will be attainable in the next few years, making it possible to construct such a score. Once a polygenic score with an R^2 of 12% can be calculated for each genotyped participant in a study, a sample of only 62 participants will be needed for 80% power to detect its effect.[70]

These researchers leverage the polygenic risk score to argue for "a shift away from candidate-gene studies" toward GWAS, a reorientation of "the focus of much research on the genetics of behavioral traits," and the creation of "new research infrastructures."[71] They mobilize their fuzzy-set methodological innovations in ways that enlarge their influence in science and put them on the map as novel messengers of the gene.

Thus, with every innovation, social genomics invests deeper in a gene-focused science that could take much more stock of the complexity of the environment. This isn't the fault of single researchers or research teams, nor is it their intention. Rather, it reflects the orthodoxy of postgenomic science. Nevertheless, in the end what we have is not a more socially and environmentally sensitive genomic science. Instead, social genomics presents society with another genes-first science that privileges genetic explanations over deep and meaningful engagements with the social and built environment.

I must state again that I don't see these moves as representing the guiding intentions of individual scientists (something you will learn more about in the following pages). It is important to see the ways that scientific collectives are shaped by the wider social and institutional context. Today there is a high premium placed on DNA evidence and evidence-based medicine from governance and the international science community. To play in the big leagues, these aspects of research need to be not only grappled with but also focalized by research teams. As such, even social science-led gene-environment (or "G by E") research will necessarily be better phrased as "genes first" science in the years to come. And sociogenomics, the paradigm for our time, will continue to be not only nature first but genes first.

CHAPTER 4

DETERMINING DIFFERENCE

My research till this point has shown that in the last decade, under the sociogenomic paradigm, we have witnessed a big move toward casting human difference in terms of genetics. Having the ability to map human genomes and trace ancestral origins has not led to a waning of folk concepts of race, gender, and sexuality. Rather, it has led to new, more robust forms of the genetic essence of them. Take race, for instance. At the advent of personal genome mapping, many scientists agreed that race was a concept better left in the folk realm.[1] They maintained that it was indicative of social relations but was not a good proxy or organizing factor for biology and biomedicine. Yet every day it seems we have new studies by race, new global projects by race, and new health standards by race.[2] And new markets are constantly opening up to produce race-targeted medicine.[3]

Does this pattern hold for the new sociogenomic science of social genomics? How is its genes-first MO impacting this wider trend? Here I'll take us through some of the major axes of difference—race, gender, and sexuality—examining each in turn. Triangulating the ways the pop media, science literature, and social genomics research community conceive of race, gender, and sexuality, I'll demonstrate that human difference is being geneticized in novel ways. As a result, a new form of essentialism, "the real, true essence of things, the invariable and fixed properties which define the 'whatness' of a given entity," is taking hold.[4]

"Blacks have double the chance . . . "

Throughout the news in this postgenomic era race appears as a given. Take one of the early reports on MAOA that set off a media storm in 2010. In "'Warrior Gene' Predicts Aggressive Behavior after Provocation," *Science Daily* reports that the low-activity MAOA gene is more prevalent in "Western" populations than in all others, stating that "only about a third of people in Western populations have the low-activity form of MAOA. By comparison, low-activity MAOA has been reported to be much more frequent (approaching two-thirds of people) in some populations that had a history of warfare."[5] This article and its many spin-offs in the leading news outlets have spurred articles and blog posts like "Whites Lowest Instance of MAOA-L Gene,"[6] which present other studies showing that "American black males," "Latinos," and "American Indians" have double the chance as "American white males" for having the gene, and that blacks are "13.5 times more likely" to possess a rare variant that is associated with extreme violence and aggression. This also has prompted readers to explore the litany of studies that have spotlighted the genetic exceptionalism of blacks, Latinos, and Pacific Islanders.

Indeed the West versus the Rest has been a common implicit way of invoking race. In "Are Our Political Beliefs Encoded in Our DNA?"[7] a 2013 *New York Times* article on genopolitics, columnist Thomas Edsall presents data on, among other research, nine studies on the genetics of political ideology in the West. This summarization, which discusses Australia, Hungary, Denmark, Sweden, and the USA as a unit, inspires one to imagine whites as cordoned off from the rest of the world. Though the research within the presented studies focuses on "within Europe" supposedly intraracial variation, the sloppy wording here broadcasts genopolitics as race specific. Later in this same article, political scientists John Alford and John Hibbing are quoted as saying that "fissures in the polity now match divisions in people's biology," including those along the lines of race, sex, and culture.[8] While Alford and Hibbings are likely referring to race, sex, and culture as a new form of identity politics, and thus as political issues, the article implies that genetic predispositions run along racial, sexual, and cultural lines.[9]

In handling behavioral findings that are stratified by race and ethnicity, news outlets often show scientists dropping racial innuendos. For example, many scientists present the sociogenomic "differential susceptibility theory," the theory that our own genes make us susceptible to environmental influences in different ways, by using examples like this:

> Some people born with the Warrior Gene and others like it have done quite well for themselves. "If you're on Wall Street, you want aggressive behavior because you want success." . . . The Wall Street banker with a dopamine deficiency—like the one found in Finnish study's subjects—would make decisions seen as more acceptable by society than his twin in the ghetto, where most would probably agree there are a different set of rules.[10]

Despite using the term "twin," it is left to readers to imagine the Finn speculating in a three-piece suit versus the darker-skinned thug committing murder in the ghetto.

Though the vast majority of sociogenomic news articles published in recent years have been guilty of spreading stigmatizing stereotypes, I'd like to share cases where coverage has done otherwise. One article that appeared in the *Wall Street Journal* in 2013 titled "The Genetic Code for Genius"[11] presents the Beijing Genome Institute's hunt for the IQ gene (a GWAS of geniuses) including voices of critics who worry that the data will be misused:

> Research into the science of intelligence has been used in the past "to target particular racial groups or individuals and delegitimize them," said Jeremy Gruber, president of the Council for Responsible Genetics, a watchdog group based in Cambridge, Mass. "I'd be very concerned that the reductionist and deterministic trends that still are very much present in the world of genetics would come to the fore in a project like this."[12]

Alas, though the blogosphere and local news media both hold lively debates about the potential racist implications of the growing slew of social genomics reports, only this article has raised a counterpoint in the major news media.

Social Genomics on Race

Unfortunately, social genomics only goes so far in helping the situation. While my conversations with social genomics researchers and analysis of social genomics publications generated hundreds of statements on race (over three hundred, to be more precise), only a handful include definitions or conceptualizations that challenge the media's status quo.

Three social genomics publications have addressed race head-on. In "Ethnicity, Body Mass, and Genome-wide Data," researchers attempt to show GWAS's utility for health disparities research by comparing GWAS results of obesity in different racial groups.[13] They find that "the genetic differences across racial and ethnic groups are evident but quite small in magnitude . . . genes do not account for observed ethnic differences in [body mass index]." The team concludes that social genomics can prove what many only speculate about with "well founded" skepticism and criticism:

> These results bolster the position of social epidemiologists who look to social context (e.g., friends, schools, workplace, residential areas) to account for important differences in the prevalence of obesity among racial and ethnic groups. Social epidemiologists have focused on normative and institutional factors that may increase the risk of obesity for racial and ethnic minorities through lifestyles and access to a healthy environment. This manifests through an emphasis on health behaviors and conditions stemming from the probability that various minority groups are likely to be of lower socioeconomic status and cultural differences between racial and ethnic groups in regard to food preferences and beliefs about a healthy body weight.[14]

This article alone provides an alternative to the media's message that race is entirely a product of one's DNA.

The second article to pointedly reflect on the meaning of race is "Data Quality Control in Social Surveys Using Genetic Info."[15] This article also offers up social genomics as a method of quality control; however, it directs researchers to use a "proportion identical by descent score [to] identify 'misreported' and unreported sibling type and detect misrepresented participants" and a "bio-ancestry score [to] repair and recover missing race and discrepancies among different measures of self-reported race." Its authors give a real example of someone multiply categorized in the Add Health dataset:

> Because this respondent self-reported "other" race at Wave I, and the interviewer reported "white" race at Wave III, it is difficult to decide which of the two variables should be put in the missing race at Wave III. In addition, for the case in which all of the self-reported races are missing (pseudo ID 1005), the missing variable might be imputed as "white" by the interviewer-reported race. But this person's bio-ancestry scores present a clearer picture, as he or she possesses about 49.9 percent European ancestry and 37.5 percent East Asian ancestry.[16]

They continue with an example from the ROOM study:

> Note that in the bottom panel, the six individuals' housing application races are not listed as "white," "black," or "East Asian." If the race information in the housing applications was not available, the race of these 13 respondents would be treated as missing. With the bio-ancestry scores, however, each of the 13 respondents could be assigned to a nonmissing race variable.[17]

This study thus offers a different form of assistance altogether. It encourages demographers and epidemiologists to use genetics to substantiate, or fill in, racial values.

In their paper "Genetic Bio-ancestry and Social Construction of Racial Classification in Social Surveys in the Contemporary United States," this team goes even further.[18] They begin with the paradox that social scientists don't take advantage of genetics in their study of race, and that biologists don't consider the social context when they measure race. They then try to solve the paradox by demonstrating the ease with which "genetic bio-ancestry" and social race can be merged into one. As they argue:

> First, we replicate the match between genetic bio-ancestry and self-reported race across a number of independent data sources (two US and two worldwide sources). . . . Second, we show in a test of the "one-drop rule" (the century-old US social and legal practice of treating individuals with any amount of African ancestry as black) that the influence of bio-ancestry on racial classification depends on how black and white are historically and socially defined. In the absence of bio-ancestry, the "one drop" cannot be measured, and thus the rule cannot be tested directly and generally.[19]

Again, this team emphasizes that race is a "more or less slippery and subjective procedure of self-identification" that can change from one study to the next or one person's assessment to the next. Yet they emphasize sociogenomic science's ability to help determine the truth, promising that "a socially influenced definition of race need not preclude any logical basis for race/ethnic classifications" given that genetics has proven there to be distinct "genetic clustering across geographically separated human populations."[20] Though these researchers conclude that "the effect of bio-ancestry depends on social, historical, and cultural context," they insist that genetics can provide a corrective to the messiness of the social assignment of race.

Aside from these three articles, most social genomics publications make rote references to racial differences without defining what they mean. For example, in "The Long Arm of Adolescence-School Health Behavior Environments," researchers say that their analysis of the 5HTTLPR gene in Wave IV Add Health samples confirms that "[Hardy Weinberg Equilibrium] was met for four of the five major racial/ethnic groups."[21] In another example, while developing and evaluating a polygenic risk score for obesity, scientists report its utility in ARIC (Atherosclerosis Risk in Communities) "whites" versus "African Americans."[22] And in the study "Evidence of Gene x Environment Interaction in Victimization Results from Longitudinal Sample of Adolescents,"[23] researchers develop a victimization scale based on "how many times in the past 12 months someone shot them, stabbed them, pulled a gun or knife on them, jumped them, or physically fought them" and then analyzed genetic associations for "the White subsample, and for the Black subsample," citing known racial differences discovered in population genetic research. In two other studies on antisocial tendencies and delinquency,[24] researchers home in on "White and Black male adolescents." All in all, a very small number of these articles explain why the authors use race, and then only in terms of population genetics.[25]

An even smaller set of articles explains in detail how researchers measure racial difference. For example, some scientists offer assortative mating as an explanation. In "Genetic and Educational Assortative Mating among US Adults,"[26] one team states that "intraethnic assortative mating among Americans of European descent is well documented and small differences in allele frequencies across European ethnic groups are easily identified with genome-wide data." They warn that without proper controls for race, "population stratification, small differences in allele frequencies that may exist across socially defined racial and ethnic groups," will continue to plague gene-environment researchers.[27] Similarly, in "Correlated Genotypes in Friendship Networks," another group of researchers justify their study by stating that "geneticists have shown that populations tend to stratify genetically, this process result[ing] from geographic sorting or assortative mating," but they have yet to determine other reasons for sorting.[28] They do not explain the social-environmental factors that lead to assortative mating, factors such as racial stratification by neighborhoods and schools, which themselves are products of redlining and redistricting.[29]

Race as Population Stratification

Indeed, researchers appear to mostly think about racial difference in terms of evolutionary continental divergences often referred to as "racial divergence." In one scientific conference, one researcher discussed the perils of population stratification, warning about racial confounding.[30] He introduced participants to ancestry informative markers, genetic markers that can be used to sift continental groups apart. Terms like "the white race/ethnicity or European population" were used in contrast to "the African American population" and "Sub-Saharan Africans," who were conflated as one race. This presenter agreed with a number of conference-goers that racial divergence seemed a problematic paradigm, and reminded them that race could become a huge methodological roadblock if not handled properly.[31]

A common strategy that study teams have adopted is to simply work with people of European descent. Researchers have scientific and ethical rationales for doing so. As one economist told me, people of European descent are easier to study because they have already been studied more.[32]

> Most of the cohorts are of European descent. So that's where we can get the biggest sample size. Also the European descent individuals are relatively well characterized. So it's known, for example, what the principal components mean and what the best ways are for addressing potential population stratification within individual samples.[33]

Another researcher said:

> The assumption is that you have a common variant structure, hence the standardization. And that's simply not the case when we are looking at genomes that come from different ancestries. Now, in some sense there are a lot of different ancestries that one could look at, and when the ancestries are relatively closely linked in terms of evolutionary time, there are some statistical methods we can use to get around those differences. So if you are dealing with Europeans, it's true that people who are sort of Northern European descent have genomes that look a little different from people who have Southern European descent. But those are differences that the statistical methods we have today can help us work around.[34]

He continued, describing population stratification in terms of "the chopstick effect":

Let's say you're a researcher and for whatever reason you are interested in study-ing the genetic underpinnings of chopsticks use, right? And so you ran one of these kind of studies that we run, those gene discovery [studies], and you would find all sorts of genes there that it looks like are really the chopsticks use. But of course we know that the mechanism involved there is just an advanced cultural mechanism whereby Asian people happen to have a culture to use the chopsticks. . . . So what we do is, to the best we can, we try to control for the particular characteristics of subpopulations and things like that. The way that we do that is by performing what's called principal component analysis, where you have a large number of variables, and what principal component analysis does is it reduces them into a very small number of variables that capture a large fraction of the variation in the original millions of variables. . . . There are actual, like, true, interesting biological things. It happens to be that one, race, has this characteristic.[35]

He expressed the prevailing belief in intraracial studies as the methodologically safest way forward, while illustrating how researchers use genomic methods to sort DNA samples into broad continental groups. With principal compo-nent analysis, researchers separate out what they see as confounding racial attributes.

In print, mentions of population stratification often coincide with racializa-tion of a trait. Researchers warn that their data can't be generalized to other racial groups in order to do their due diligence, but this ends up reinforcing the notion that races are discrete and mutually opposing entities. For example, in "Monoamine Oxidase A Gene (MAOA) Predicts Behavioral Aggression Fol-lowing Provocation," scientists stipulate that their data on "MAOA and Ag-gression" collected in "Caucasian adolescents only . . . may not generalize to populations of differing ethnicities and cultural norms." Then they point to studies in other populations, such as one on Pacific Islander Maoris, which display "about a 60 to 65 percent incidence of the MAOA-L polymorphism whereas Western European populations typically report about a 30 to 35 per-cent rate of appearance."[36] Or as another study on risky behavior states:

Our samples are very homogenous in terms of race/ethnicity. Although this ho-mogeneity reduces the bias in instrument effects as a result of population stratifi-cation, it may limit the generalizability of the study results to other populations. This limitation highlights the need to study the utility of these genetic variants as instruments in other populations and samples. However, it is important to

recognize the potential bias resulting from population stratification when studying diverse populations, such as those found in the United States. If the allelic
distributions of the genetic instruments vary by ancestry, the IV estimates may
be biased. Furthermore, direct adjustment for self-reported measures of race
and ethnicity may be inadequate to remove this bias, as these may not effectively
account for differences in allelic frequency by ancestry.[37]

While the authors' statements about generalizability are valid and accurate,
they don't change the fact that their study design inadvertently reinforces a
sense of essential racial difference. As I have elsewhere remarked many times
before, using a continental system for managing population stratification like
ancestry informative markers only gives validity to unwarranted biological notions of race.

An expert on postgenomic population science put it best when she said
importing principal components analysis doesn't help researchers get around
the problem of race:

> Number one, it accepts the premise that there is enough population structural
> variation that you would have to do that; but number two, it doesn't avoid the
> idea that there are caste-like structures within a population—so, long histories
> of endogamous kinship among people from different economic classes, even
> within the European, whatever that is, population. It's interesting, because it
> does show their sensitivity to issues of race-based group behavioral claims. The
> idea that they will be able to forgo forever, or even the kind of idea that what we
> will end up here with is a different set of markers for behaviors for each conti-
> nental ancestry, neither of those themes—well, the first one doesn't seem plau-
> sible. They won't be able to put this off forever. The second one is also a concern.
> It doesn't avoid the idea of difference.[38]

As this scientist remarked, though researchers seem to, as they often put it,
"desire to avoid things like racial stigma," their study design promises "future
dilemmas."

This Doesn't Apply to Girls

Far more objective seeming and rigidly stereotypical than depictions of race is
how gender is portrayed in the sociogenomic news. Gendered claims about the
genetics of behavior populate the media, inflecting articles with content that

can feed into sexism and discriminatory ways of thinking. Unlike with race, when they use the terms "Western" or "West" in place of "whites" or "people of European descent," broadcasts make no attempts to euphemize the binary categories of male/female, men/women, and masculine/feminine. Far from it, these binaries are presented as completely natural and more genetic than readers could have previously known.

A typical claim about the gendered nature of genetics looks something like this statement made in the *Times* of London in a 2009 article called "Gang Genes":

> An American criminologist claims that whether or not a boy joins a gang is substantially influenced by his genes. Not only that, but the level of violence used by individual gang members can also be predicted courtesy of the same gene. . . . Dr. Beaver, who studied the DNA data and lifestyle information of 2,500 adolescents, said: "We also found that variants of this gene could distinguish gang members who were markedly more likely to behave violently and use weapons from members who were less likely to do either." The same finding does not apply to girls.[39]

Meanwhile, there are statements like this one taken from a 2010 *Economist* article called "The Biology of Business Management":

> Dr. Shane observes genetic influence over which jobs people choose (see chart), how satisfied they are with those jobs, how frequently they change jobs, how important work is to them and how well they perform (or strictly speaking, how poorly: genes account for over a third of variation between individuals in "censured job performance," a measure that incorporates reprimands, probation and performance-related firings). Salary also depends on DNA. Around 40% of the variation between people's incomes is attributable to genetics. Genes do not, however, operate in isolation. . . . Previous research had shown that people exhibiting personality traits like sensation-seeking are more likely to become entrepreneurs than their less outgoing and more level-headed peers. Dr. Arvey and his colleagues found the same effect for extroversion (of which sensation-seeking is but one facet). There was, however, an interesting twist. Their study of 1,285 pairs of identical twins and 849 pairs of same-sex fraternal ones suggests that genes help explain extroversion only in women. In men, this trait is instilled environmentally. Businesswomen, it seems, are born. But businessmen are made.[40]

In articles like these, reports lay out a genetics of gender that feeds folk beliefs that men and women are interminably distinct down to their cellular core.

Coverage of genes has been so gendered in the media that it would appear that genes themselves are innately gendered—in other words, that they only appear in men or women, not both. In 2012, *Good Morning America* ran a program on motherhood called "Who Has the Mom Gene?"[41] The show opened with reporter Juju Chang asking viewers, "You know, have you ever wondered why some women hear their biological clocks so loudly, and others, not so much? Well those researchers at Rockefeller University suggest maybe the maternal instinct is just that, it's an instinct linked to a specific gene they've isolated." Then it cut to Chang interviewing a five-year-old girl named Saorise, who like her sister, Rory, expressed a desire to be a mother one day:

> Do you know for sure that you want to be a mommy?
> Yes.
> And how many babies do you want?
> Two.

For the rest of the program Chang interviewed mothers and parenting experts who agreed that "the longing Saorise and Rory feel, to play with strollers, and feed and nurture baby dolls from such an early age" was something they likely were born with. As one mom said, "I can always remember playing with dolls, and always thinking, you know, I'm gonna be a mom." A parenting expert told future moms not to worry, that they could "compensate" for their true feelings about being a mom if the "innate pull isn't there." Viewers were left with the takeaway that intensive parenting is something bred in women, not men.[42]

Perhaps the most covered gene in the media, the "warrior gene," is the best example of an explicitly gendered gene. Coverage of this gene includes reports like "'Warrior Gene' Reported Rife among Young Thugs"[43] and "'Warrior Gene' Linked to Gang Membership, Weapon Use,"[44] which have made statements like, "The findings apply only to males; girls with the same genetic peculiarity seem resistant to its potentially violent effects." While articles often explain that males have two copies of the gene in question, and thus are more likely to suffer from problems related to the gene, they do nothing to raise the issue of male socialization to be aggressive, dominant people in society. Even those that mention the role of the environment (which still is reserved for discussing family upbringing and not peer groups, neighborhoods, and media influences) only do so insofar as to raise the issue of better child-raising strategies for troubled kids.

It's important to note that researchers do try to complicate the picture. But even in cases where those responsible for the study try to dial down the stereotyping, determinism reigns. Take this 2014 *Washington Post* article, "Are There Genes That Lead Men to Enjoy Fighting?" which says that "some people, especially some men, like to fight. They even like to kill."[45] It quotes study leader Rose McDermott as saying:

> To be clear, I am not, nor will I ever, claim that there is a gene for anything; such a construction betrays a deep misunderstanding of the nature of genetic pleiotropy, whereby it requires many genes in combination and in interaction with the environment, along a long trajectory of developmental processes, to produce any behavior.[46]

Yet despite her best efforts, the article goes on to talk about the critical role of this gene in striking up in certain environments to the point of making whole cultures festering sites of violence.

Gendered claims about the genetics of social behavior abound in the media to the point where they infuse articles that aren't necessarily reporting specific study findings. As exemplified in "A Cure for Character," an article on innate personality flaws, reports are quick to use genetic claims to bolster gender theories. "A Cure for Character" quotes famed political scientist James Wilson, who states:

> The part of the brain that stimulates anger and aggression is larger in men than in women, and the part that restrains anger is smaller in men than in women. "Men," Wilson writes, "by no choice of their own, are far more prone to violence and far less capable of self-restraint than women."[47]

This statement only reinforces the notion that it's simple nature for men to be one way and women to be another.

Many articles making gendered claims report genetic reasons for disparities even when none have yet been found. Another article on antisocial behavior in "callous-unemotional children" reports:

> The trait appeared to be strongly influenced by the genes parents passed on to their sons. In fact, about 80 percent of the likelihood of boys developing emotional problems were explained by inheritance. But in the much smaller proportion of girls in the study who lacked normal empathy, inheritance played almost no role.[48]

While the scientists on the study called the gender disparity "puzzling," even though the study included "too few emotionally disturbed girls" to make a definitive assessment, they took no time in conjecturing that the gene must be on a sex chromosome.

As with race, there have been nearly no reflexive media articles on the genetics of gender. The one and only article presenting dissenting opinions on the genetics of gender is a 2012 *New York Times* article titled "Scientists to Seek Clues to Violence in Genome of Gunman in Newtown Connecticut." This article states:

> Some researchers, like Dr. Arthur Beaudet, a professor at the Baylor College of Medicine and the chairman of its department of molecular and human genetics, applaud the effort. He believes that the acts committed by men like Mr. Lanza and the gunmen in other rampages in recent years at Columbine High School and in Aurora, Colo., in Norway, in Tucson and at Virginia Tech are so far off the charts of normal behavior that there must be genetic changes driving them. We can't afford not to do this research, Dr. Beaudet said. Other scientists are not so sure. They worry that this research could eventually stigmatize people who have never committed a crime but who turn out to have a genetic aberration also found in a mass murder. Everything known about mental illness, these skeptics say, argues that there are likely to be hundreds of genes involved in extreme violent behavior, not to mention a variety of environmental influences, and that all of these factors can interact in complex and unpredictable ways.[49]

This article quotes leading neurogeneticist Robert Green of Harvard Medical School as saying that it's "almost inconceivable that there is a common genetic factor to be found in mass murders." Finally, it reminds readers of the sociopolitical stakes in asserting genetics as the cause, noting the history of eugenics, XYY portrayals, and other scientific blunders that have led to devastating consequences for people all over the world.[50] However, even this lone critique does not directly address the gender bias that permeates media portrayals.

Social Genomics on Gender

Far from the careful and almost proactive way that social genomics researchers deal with race, gender is barely given a thought. There have been hundreds more invocations of gender differences in the popular media, science media, and at science conferences and meetings with the public than there have been of race. Yet there have been nearly no explicit mentions of it in publication, let alone attempts to engage it critically.

Across the science media, social genomics studies cite gender in a dry, uncritical fashion. For example, "A Test of Biological and Behavioral Explanations for Gender Differences in Telomere Length" mindlessly mentions "women's midlife advantage in LTL [leukocyte telomere length]" and "the narrowing of the gender gap in telomere length during the post-reproductive years," explaining it by reference to hormonal differences between the sexes. Gender is used as a metonym for sex.[51] In "The Nature-Nurture Debate Is Over, and Both Sides Lost: Implications for Understanding Gender Differences in Religiosity," researchers also let hormonal references fly. They write that "different chromosomes, variable levels of hormones such as testosterone and estrogen, etc. manifest themselves in many different social outcomes, including aggressiveness, dominance, nurturance, mating behavior, sociality, parent-child bonds, and risk aversion, among many others."[52] They also cite "significant genetic effects" on family relations like warmth, empathy, disputes, anger, monitoring, knowledge of activities, romance, and more.[53]

Some articles go further than rote citing to instead advance more pointed essentialist claims about gender. "Prenatal Smoking and Genetic Risk" argues for the existence of gendered gene-environments:

> Our results provide evidence that the influence of genetic risk factors on behavioral problems were most pronounced for children exposed to prenatal smoking. In other words, genetic risk factors interacted with prenatal smoking to predict the early emergence of EBP [externalizing behavioral problems]. A fourth, and related point of interest, was that the interaction between prenatal smoke exposure and genetic factors appeared to be isolated to females. . . . Our results suggest that the effects of prenatal smoking and genetic risk factors may coalesce in females to magnify the influence of genes.[54]

These scientists call for more studies to replicate their results, to let "gendered interactions emerge" and to "elucidate the gender gap that exists for aggressive and antisocial tendencies." Again, here female and male are simply biological entities, with their own genetic-hormonal profiles.[55] There is no acknowledgment of social circumstances informed by experiences of stratification or categorization.

One example in which researchers try to shy away from blanket determinisms shows just how often they end up reinforcing a biological concept at the expense of a well-developed social one. "In Defense of Genopolitics," for example, warns:

> A wide range of political phenomena are influenced by a person's sex, and although we do not claim that these effects are all purely genetic, neither do we doubt a reported association between sex and one political behavior merely because it also predicts other political behaviors. Sex can influence specific hormonal responses, but it can also induce gender-specific roles in the self and others. Each of these could be considered an endophenotype with many phenotypic consequences.[56]

The causal chain here starts with genetics, moving forward to hormonal responses and gender-specific roles, and on to the phenotypes in question which are social outcomes that possess gender gaps. There is no acknowledgment of social outcomes informing gender roles, and gender roles informing biology.

Many articles that advance arguments about gender specifically lean on evolutionary theories to bolster their claims. "The Influence of Three Genes on Whether Adolescents Use Contraception" reminds readers that "the potential consequences of unprotected sexual intercourse differ sharply for the two sexes."[57] Its authors state that females may get pregnant, and thereby change their life course with childbearing and rearing, while males need not worry. These "disincentives" may modulate genetic propensities. What's more, "males of most mammalian species show a stronger desire towards variety in sexual partners than do females." The authors conclude that the DAT1, DRD2, and MAOA genes "have a role in regulating neurotransmitter function, and thereby contraceptive use, among adolescents and young adults," which they found to play out differently in females and males, and which must be moderated differently by these varying evolutionary roles.[58]

Such depictions can get very technical with their presentation of evolutionary language. Take "Violating the Monotonicity Condition for Instrumental Variable: Dimorphic Patterns of Gene-Behavior Association,"[59] an article which starts by warning readers that "many gender differences in behavior and disease profiles are not caused by females and males having different genes." Instead of pointing to environmental causes, its author argues that dimorphisms "manifest in the occipital cortex" and then are "conserved beyond species: human and old world monkeys share the same sexual dimorphic pattern," therefore "gender differentiated expression is evolutionarily advantageous and functioning."[60] Likewise, in research on resiliency to victimization, researchers characterize male predispositions to violence and aggression as traits "highly advantageous in the evolutionary past":

> In the evolutionary past, it is easy to see that violent and aggressive behaviors (many of which would be criminalized today) would be beneficial in that the most violent persons, in comparison with the most passive persons, would tend to live longer, produce more children, and perhaps even have more of their children reach the reproductive years. The reason for this advantage is relatively straightforward: Those who are adept at using violence are better able to protect their kin, to protect themselves, and the use of violence may even be one characteristic of female mate choice.[61]

Interestingly, in another article several of these authors attempt to explain the gender gap in parenting and adolescent self-regulation, and what they call "genetic plasticity" more broadly. While they leave open the possibility for females to be more susceptible to parenting, they couch the difference in talk of differential developmental biology as a result of evolutionary dimorphism. Like the hundreds of publications merely mentioning gender, they thus end up reducing it to biological processes.

So where are the publications that speak on what gender actually is and how to deal with it? Remember when one team of researchers said that genetic data could be used as a corrective for social categorizations of race?[62] Theirs is also the only study to do this for gender. They say that "sex chromosomal information may help cross-check self-reported sex":

> XX and XY are the common combinations of sex chromosomes for females and males, respectively, in humans. The gonads need to be formed and differentiated for humans to develop sexually. The Y chromosome induces testis formation and is the dominant determinant for male development. Without a Y chromosome, gonads differentiate into ovaries, resulting in female development. On the human Y chromosome, the sex-determining region (SRY) plays the crucial role. The SRY is a single-exon gene. This gene contains transcription initiation sites, and these transcripts are identified in adult testes and other male tissues. The SRY initiates testis development from early bipotential gonads. The complex sex determination process also involves genes on autosomes. For example, DMRT1, which is located on chromosome 9, has been found to be associated with haploinsufficiency in the form of XY sex reversal. Different combinations of the sex chromosomes other than XX and XY exist in the human population. XXY or XXYY males, individuals with the condition known as Klinefelter

syndrome, have two X chromosomes and one or two Y chromosomes. Persons with Turner syndrome have only one sex chromosome; that is, 45 chromosomes in total, including only one X chromosome. The proportion of non-XX females and non-XY males is considerable. About 1 out of 500 to 1,000 boys are born with Klinefelter syndrome, and 1 of every 2,700 live births results in a child with Turner syndrome.[63]

These researchers give two caveats about using sex genetics to control for survey research inadequacies. First, they note that "it is impossible to be 100 percent certain of an individual's sex by only looking at the sex chromosomes," but they follow by saying that "given the relatively low prevalence of unusual combination forms of sex chromosomes, the common forms of XX for females and XY for males may still serve as a reference." And as with race, they provide a sociological caveat that gender is "socially and biologically constructed in complex ways. . . . Individuals enact and reinforce their femininity or masculinity in social interactions." But despite the nod to social processes of gender, these scientists conclude that all in all, DNA data can be that illusive corrective to gender multiplicity.[64]

X versus Y

As with researchers' reliance on the genomics gold standard of population stratification correction for managing race in study design, they use the genomics gold standard of managing sex difference when dealing with gender in study design; that is, X versus Y sex stratification, or separating out males and females in analytical runs of the data. As one researcher told me, "men will, just by definition, have fewer number of alleles in your study. And so comparing results by gender would be just not the same." She said she compares "same sex" on the sibling pairs she studies. Another said he and his team "always include gender as a control variable. In most cases, we also will run separate models on males and females; so the X sets as independent and then we run the same model on each independently."[65] Still another study leader said:

> We're doing gender-stratified analysis. But it's not really gender studies per se. It's not that we're interested in gender differences per se; it's just that for most of the things that we're interested in you cannot really exclude a priori that the same genes are going to have the same effect for males and for females.

This researcher explained that "males and females face different environmental constraints, so it may be that the effects are actually mediated by gender."[66] This was the closest I heard of a complex gene-environment definition of the situation, and it still left many questions about what constraints are unanswered.

While most researchers work at times on X-linked traits, a small number don't. I found that even these individuals who don't need to stratify by sex push for sex-stratified studies to deal with gender gaps in social outcomes. As one researcher who does a lot of work in education put it: "We occasionally have dealt with men and women separately, though there's never really an issue."[67] Another researcher, who works primarily on financial risk, said that "splitting samples by sex or thinking about sex-specific things" was warranted when you are dealing with "sex-linked phenotypes."[68] In other words, gender-stratified phenotypes can be understood best with sex-stratified genotyping.

Even researchers who have only worked on outcomes like addiction and obesity, which are prevalent in both genders, aver the irrefutable genetic difference of men and women, and their need to be treated differently. As one researcher told me:

> One of the things that genetics makes unequivocally clear is that certain social categories have biological correlates. That doesn't mean that the differences, the phenotypic differences between the social categories that we observe, are caused by those biological correlates, but boys and girls have different chromosomes. And frankly, we know tons about sex differences in human physiology. I mean there clearly is a biological basis of sex, just as clearly as there is a biological basis of race. It's not the only source of variation; race represents a lot more than biology. But people are different sizes, they have different skin colors; those are genetic characteristics.[69]

For this researcher and others, sex differences must be attended to at the level of biology and thus study design. Ignoring that is tantamount to doing bad science.

This isn't to say that I didn't hear the occasional mention of gender disparities and their social origins. One male researcher told me:

> I've seen crosstabs that I would not publish because I would think about some of the social implications. Or they don't make sense, and so I will put those to the side. I call it racial politics and gender politics. I use the word "politics" loosely, but [mean] the potential issues that it raises. I mean if we find something—for

example, it's on the X chromosome—that women tend to have, that relates to intelligence, it gets into these gender and intelligence debates.[70]

This researcher said that publishing work that could have sexist repercussions only "tends to harm the science more than anything else." He didn't want to jeopardize the nascent science, or make a bad name for sociogenomic sciences writ large.

And finally, one female researcher spoke out about institutional sexism, saying:

> I have some junior people come to me to ask me for advice, and my whole thing is like, "Don't become an academic," you know, especially if you're a woman, es-pecially if you want children. I think that it's very hard in good faith to encour-age them. I do, but I'm pretty discouraged. Psychology shows us, social science progresses one female at a time. I think that's my cynical side, but it's so true. Like you have to teach below you, not above you, right? If it's really important, then the people who are threatened by what you're doing, that's not your audi-ence. You don't want to convince people older than you. You want to convince people younger than you. I can only hope that those people will be more inter-pretive, more creative in a way that structures their institutions. I'm hoping that because of the influence of the World Wide Web and what it does to encourage interconnectivity across things, that maybe they'll be able to do it. But disciplin-ary boundaries within established institutions are so strong, especially like old dead wood that has been there for forty years and doesn't want to be changed or threatened. It's just very frustrating. *Very* hard.[71]

This researcher was the only person to bring up sociological factors of sexism and patriarchy (and only one of two to mention sexism at all in any meaningful way). Yet no one discussed sexism or patriarchy in terms of the gene-environment re-lationship. As in their publications, these issues were not a party to researchers' casual explanations of what gene-environment interactions are all about.

"A large study of gay brothers shows . . ."

Sexuality is depicted in mixed ways in the sociogenomic media. When articles are about social outcomes that are already stigmatized with a certain gender, like rape, reports are cut-and-dried about their genetic claims. When the news is instead about things that are considered to be male and female traits, like homosexuality, reporters tread very carefully around their biological roots, avoiding blanket statements and representing diverse voices on the matter.

A recent media uproar around gay genes found in brothers shows just how balanced portrayals have been regarding the latter. In 2014, the Associated Press wrote:

> A large study of gay brothers adds to evidence that genes influence men's chances of being homosexual, but the results aren't strong enough to prove it. Some scientists believe several genes might affect sexual orientation. Researchers who led the new study of nearly 800 gay brothers say their results bolster previous evidence pointing to genes on the X chromosome. They also found evidence of influence from a gene or genes on a different chromosome. But the study doesn't identify which of hundreds of genes located in either place might be involved. Smaller studies seeking genetic links to homosexuality have had mixed results.[72]

This news blast, which flashed throughout the international news media, paired headlines like "Large Study of Gay Brothers Suggests Genetic Link for Male Homosexuality" and "Genetic Link for Male Homosexuals" with softer secondary headlines like "The research 'is not proof but it's a pretty good indication'"[73] and "Experts, however, are skeptical." These articles cover critical voices in depth and with as long a treatment as study leaders' voices. They also present critical voices of people in greater positions of power within the sciences, such as former president of the American Society of Human Genetics Neil Risch.

Coverage of sexual orientation has also been balanced in another way. Reports have highlighted the political stakes findings may have for people. For example, the *Guardian*'s 2015 article "Born This Way? Society, Sexuality and the Search for the 'Gay Gene'" presents research to counter the myth of there being a "gay gene" and then concludes that even if sexuality isn't in our genes, gays and lesbians deserve equal rights:

> Queer relationships should be embraced, not because homosexuality is genetic, but simply because there is nothing wrong with them. While gay gene arguments may seem like a way to push the rights agenda forward it can actually have the opposite effect limiting the debate solely to those traits and behaviours seen as genetic.[74]

A *New Scientist* article titled "Gay Brothers and Gay Genes" interprets the study as ushering in helpful proof that sexuality is in the genes: "The finding is an important contribution to mounting evidence that being gay is biologically determined rather than a lifestyle choice. In some countries, such as Uganda, being gay is still criminalised, and some religious groups believe that gay people can be 'treated' to make them straight."[75] Reports thus encourage readers to think about the social and political implications of science, and to be on the lookout for more studies that may confirm or disconfirm current findings.

Even titles of sexual orientation articles have put the brakes on knee-jerk interpretations. The *Telegraph* reported the gay brothers study with the headline "Being Homosexual Is Only Partly Due to Gay Gene" and the secondary headline "Study finds that while gay men share similar genetic make-up, it only accounts for 40 per cent of chance of a man being homosexual." In addition to foregrounding hesitance, the article, which largely featured study leaders' views, quoted them as saying:

> Our findings suggest there may be genes at play—we found evidence for two sets that affect whether a man is gay or straight. But it is not completely determinative; there are certainly other environmental factors involved. The study shows that there are genes involved in male sexual orientation. Although this could one day lead to a pre-natal test for male sexual orientation, it would not be very accurate, as there are other factors that can influence the outcome. . . . We don't think genetics is the whole story. It's not. We have a gene that contributes to homosexuality but you could say it is linked to heterosexuality. It is the variation.[76]

The opposite kind of treatment appears when we look at news on pathological sexuality. In 2015, the media was once again abuzz with reports of a study on the genetics of sex offending. BBC Radio was first to the punch with its program "Sexual Offending 'Runs in Families,'"[77] which announced that researchers had found a genetic "link" that was responsible for 30–50 percent of sexual behavior. BBC asked if behavior was more about our genes than we realized, to which the study's lead scientist argued that hypersexuality and other high-risk behaviors could very well be "genetically determined."[78] Soon after, international news outlets like *Newsweek* and the *Telegraph* reported definitive statements such as, "Male relatives of sexual offenders are up to five times more likely to commit similar offences, researchers have found,"[79] and "Genes play a leading role in the link, according to the study of 21,566 men—suggesting that certain individuals are born with a propensity to commit sexual offences."[80] Media accounts also inflated the statistics saying that "statistical analysis indicated that

40% to 50% of the differences in risk seen between close relatives of offenders and men from the general population were genetically driven."

Reports from ABC and CNN have since provided more detail about the study, noting that "the biological sons of a sex offender were five times more likely than the norm to commit sex offenses"[81] and that "maternal half-brothers were only twice as likely to do so."[82] In these broadcasts, the study's leader has been quoted as saying: "There is a genetic component. . . . We shouldn't discard the genetic component. It isn't all about environment." She also suggests that knowing the genes responsible for sexuality could lead to better policy for helping families with inborn pathologies, thus giving the impression that the genetic findings are insurmountable, and thus ready to be acted upon.

Notably absent from the many news blasts that have taken place around the world in venues like CNN, the *Wall Street Journal*, ABC News, the *Washington Post*, NPR, and BBC are countervailing voices. Only study scientists have been interviewed and quoted on ways of reading the data. As a result, the only statements reminding viewers, listeners, and readers of the role of the environment have been made by study scientists, who push against thinking in environmental terms. Some examples we can look to are, first, this quote by another study scientist:

> It's important to remember that it's nothing mystic. . . . People get worried about the fact that there's a strong genetic component in problematic human behaviour. Of course, you don't inherit in some kind of automatised robotic way so that you will grow up to be a sexual offender.[83]

And this one by the previous study leader:

> At the moment genetic factors are typically ignored when it comes to making risk assessments of those at high risk of sexual offending. . . . Many of the families we are talking about may already be known to social services for other reasons, and if we can predict those at high risk of offending with greater accuracy then it may be possible to shape these interventions and target education and preventative therapies where they could do the most good.[84]

In fact, the only countering scientific interpretations have been made in the science media, like in *Science*:

> Most studies point to early life experiences, such as childhood abuse, as the most important risk factor for becoming a perpetrator of abuse in adulthood.

The new study, however, did not include any detail about the convicted sex criminals' early life exposure to abuse. . . . It's extremely difficult to collect sufficient data about sexual offenders and their families to detect statistically robust patterns. Sweden is unusual because its nationwide Multi-Generation Register allows researchers to mine not only anonymized criminal records, but also to link them with offenders' family records as well. Even with access to a nationwide database, Seena Fazel, of the University of Oxford in the United Kingdom, and colleagues had to include a very diverse range of offenses, from rape to possession of child pornography and indecent exposure, to maintain a large sample size. The team did do some analysis by type of offense, separating rape from child molestation, for example. But some researchers worry that attributing a genetic basis to such a wide swath of behaviors is premature. There are also problems with relying on conviction records: Many more sexual crimes are committed than reported, and the proportion of those that go to trial is even smaller.[85]

International news outlets have not included opposing views.

In addition to matters of sexual orientation and criminal behavior, the news has devoted a lot of coverage to relationships. As with sexual pathology, the news featuring stories on the genes for relationship outcomes reports sexuality point-blank with no mention of attenuating factors. One outcome that has loomed large is fidelity. For example, the *New York Times* article "Infidelity Lurks in Your Genes" says:

> We have long known that men have a genetic, evolutionary impulse to cheat, because that increases the odds of having more of their offspring in the world. Now there is intriguing new research showing that some women, too, are biologically inclined to wander, although not for clear evolutionary benefits. Women who carry certain variants of the vasopressin receptor gene are much more likely to engage in "extra pair bonding," the scientific euphemism for sexual infidelity.[86]

This longest of reports on the research gives immensely detailed information on the evolutionary reasons for monogamy and polygamy, pair bonding and cheating, lending credence to the notion that sexuality is deep in our blood.

Reports also herald the discovery of the "Singleton gene," a gene that lowers serotonin levels in carriers and thus, as one article puts it, affects the ability for people to feel happy, feel comfortable in close romantic relationships, and therefore build and sustain them. *SF Gate* reports:

Scientists from the university tested hair samples from almost 600 students to analyze the gene they're calling 5-HTA1, which comes in two variations. People with the "G" version of the gene were much more likely to be single than those with the "C" version. Most significantly, it's reported that the connection between carriers of the gene and their single status could not be explained by other factors like income or appearance. The final conclusion of the study is that there is now "evidence for genetic contribution to social relationships in certain contexts."[87]

Again, no countervailing views or alternate interpretations are offered.

Genetic tests have been the subject of the vast majority of news coverage in the area, and most of these reports have been on TV. In 2014, for example, *Nightline* ran a program that administered "Instant Chemistry" tests to spouses. As one woman later said:

As a strong believer that "things are meant to be," I was at first skeptical that science is what brought Taso and I together. How could that be? . . . As we patiently waited for the results to be delivered, I couldn't help but to constantly have the topic of genetic compatibility on my mind. What if Taso and I lacked in having complementary genes? Would we lose our physical attraction to each other and begin to question our relationship? Will our long-term relationship satisfaction begin to dwindle?. . . . And all I can say is wow! Taso and I are 98 percent genetically compatible and, according to Instant Chemistry, this is extremely rare and only occurs in less than 10 percent of the population. We are, in fact, an outstanding match. As you can imagine, all my worries quickly vanished and a huge smile quickly formed and remained embedded on my face.[88]

Like a number of other test takers, she sung all-out enthusiasm for tests:

After going through this fantastic experience and learning more about ourselves and each other, I can truly say that I am happy I decided to take the test. Science is amazing. The more we learn about ourselves and overall genetic makeup, the better we will efficiently grow as people.[89]

Test makers, of course, have only reinforced the notion that their tests are airtight. They also insist on the simplicity of tests, as this representative of Singld-Out shows:

All you do is you spit in the cup, send the saliva back to the lab where we do our analysis, register your kit online, and do our psychological assessment. . . . I think this is going to create another solution for [people] going online and finding their ideal match. This is really the primary thing: You go read everything about this person online and then you go out on a date and you find that there's no chemistry whatsoever. There's nothing that you like about this person, so you become disheartened. You become really discouraged with online dating. So what we're doing is we're bringing the scientific methods to really tell you, "look, this person fits what your preferences are online," and then on top of that, we add another layer, which is the scientific solution that tells you that you also have a genetic compatibility with this person at 90%.[90]

Or, as this representative from ScientificMatch says, "Just a cotton swab you rub on the inside of your mouth for a few seconds" and then you will know whether you have physical and sexual chemistry.[91] Again, these reports only feature test makers and scientists citing evolutionary support for the indubitable genetic basis of sexuality, dangling promises for "healthier children and more satisfying sex lives."[92] They do not portray information about the role of the environment or discuss social ramifications. Expert proponents, and expert proponents alone, hold sway.

Just like with race and gender, I could find only one lone voice of dissent in a single news article regarding the sociogenomics of sexuality. The Associated Press article "DNA Same Couple Matching" quotes the medical director of the General Genetics Clinic at the Cleveland Clinic, Rocio Moran, as saying that genetic claims about sexuality are "ridiculous" and that test makers are just "trying to make a buck."[93] Other scientists covered say that studies do not bear up these claims, and that other physical, personality, and social factors are more important. Yet even this one article ends its critical portrayal by quoting a scientist as saying that genetics are only part of the puzzle, thus still giving credence to the resolute importance of genetics in sexuality.

Social Genomics on Sexuality

As much as sexuality factors into the media, it is all but absent from social genomic literature. In my review of its large and fast-growing body of studies, I saw almost no explicit definitions of sexuality and very few direct mentions of it. That isn't to say that there aren't a lot of articles on the topic. Articles with titles like "Age at First Sexual Intercourse, Genes, and Social Context"[94] and "Life-Course Persistent Offenders and the Propensity to Commit Sexual Assault"[95] all drill down on the topic of sexuality. But they discuss it without providing a theory of it. The first of these two articles, for instance, only defines methodology in an objective way, for example stating, "When we say sexual intercourse, we mean when a male inserts his penis into a female's vagina," thus leaving intact common assumptions that sexual intercourse is a specific sliver of heterosexual interaction.[96] The second article similarly states, "Participants were asked about items related to sexual involvement, marital status, and instances of contact with the criminal justice system. . . . In line with prior research on LCP [life-course persistent] offending, only male participants were included [in] the final analytical sample."[97] Thus, a nature-first way of portraying sexuality is most implicit in the literature.

I was able to see this trend in effect at the conferences I attended that were oriented toward the greater social science community. At one sociological conference panel on integrating sex biology into social science research, a speaker declared that sexuality and gender identity begins in utero, citing research on the INAH3 gene and its variations in prenatal girls who transitioned to become boys.[98] This researcher also talked about genetics as manifested in gonads, hormones, and the brain, which she set against culture, experience, and interactions. Another said that while genetics hadn't been proven to determine things like how someone would score on a math test, how much they'd earn, or the exact number of sex partners they would have in their lifetime, it was indubitably involved in sexual orientation and sex differences in health and disease. Both researchers cited animal model studies in mice and rats that had shown the biological origins of sexual behavior including sexual orientation, receptivity, and proceptivity.

Still, there have been a few articles wherein sexuality has been treated with a bit more concern. These articles nod to the plasticity of sexual orientation. Some claim that the social genomics of sexuality will enlighten the science community and public about heterosexism. For example, the article on the social construction of racial classification in social surveys mentioned before invokes sexual orientation as one of the main axes of identity in the United States, and it calls for researchers to really explain in detail how identities are constructed from collective beliefs.[99] However, the article also concludes that genetics is an important missing piece of the puzzle, the corrective to "an oversimplified and flattened identity account."

Another article goes further. "Learning to Love Animal Models or How Not to Study Genes as a Social Scientist"[100] reminds those interested in pursuing gene-environment interactions research that failures to replicate can come too far on the heels of sensational findings. It states:

One notable example can be found in the so-called "gay gene." Hamer et al. pub-lished an article in *Science* showing an association between a microsatellite on the X-chromosome (called Xq28) and homosexuality in men. The conclusion rested on the greater propensity of gay brothers to share genetic markers at this locus as well as pedigree analysis that showed a greater likelihood of gay men to have other gay male relatives on their maternal side (since the X that males receive always comes from their mother). Later work failed to replicate the find-ings among a similar sample of Canadian brothers and a heated debate ensued. Hamer et al.'s study is among the better of the associational studies given its ped-igree-based analysis, but like many others in the field it relies on a small, non-representative sample and purports to explain a complicated phenotype: *stated* sexual orientation. I underline "stated" for a reason: Even if the results could be routinely replicated, it may be the case that the Xq28 locus is associated with willingness to reveal homosexuality to survey takers rather than to homosexual-ity itself, given its sometimes stigmatizing status in North American culture.[101]

Here sexuality is acknowledged as a hot-button issue that can stigmatize; thus gene-environment researchers should be extra careful about their claims.

Finally, "The Genetics of Politics: Discovery, Challenges, and Progress"[102] asks whether genetics can advance public policy. Its authors answer:

Public policy is to political science what clinical studies are to the life sciences: the immediate application of research to improve human conditions. A novel stream of research is now examining how social structures and political institutions affect the environment in ways that trigger or suppress the expression of particular ge-netic factors, as well as how genetic information might shape and develop policy intervention. . . . Perhaps the most successful application to policy, although in-direct, resides on an issue central to current political discourse: discrimination against homosexuals. Although the specific results of the study remain debated, after the team at the National Cancer Institute implicated Xq28 on the X chro-mosome in male homosexuality, the concept of sexual preference began to shift public discourse from morality and choice to inherent disposition. Many factors contributed to change in attitude about homosexuality, but genetic research had an important role in shifting elite and legal discourse, which has filtered down and influenced public opinion and policies on the legality of gay marriage. Turning the eugenics movement on its head, the integration of genetics and public policy has been used to help protect individuals in meaningful ways, thereby reducing health risks, promoting healthy lifestyles, and increasing tolerance for differences.[103]

According to these scientists, "Genetic research is beginning to influence, inform, and enlighten the public, including the formulation and evaluation of significant public policies." Social genomics thus stands to improve society at the policy level in the area of sexuality.

I encountered this rationale in my private conversations with social genomics researchers. One comment I heard along these lines was:

> Even when Hamer came out in '93, saying he identified potential genetic mechanisms behind homosexuality, it turns out he was wrong. That research didn't hold up to the test of time. But the idea was partially responsible to changes in legislation saying you can't discriminate against people for sexuality. And it's the concept that it could be partially genetic, that it could have something to do with your biology to kind of move people away from what they consider to be a choice of morality to just human nature.[104]

This researcher went on to link geneticization of sexuality traits to better gender policies as well:

> We are already starting to see research that talks about differences in the need for recess between boys and girls, and that we used to introduce or had certain things that hindered women from having the best education possible. Now we are starting to kind of go the other way. And schools that are limiting recess, they are actually seeing test scores go down in boys and a lot of that has to do with your anatomy and physiology. And so if we start to see policy informed by just science itself, I think that's where—if there is any kind of equivalent to translational medicine in social science and genetics—that's where we will see it. We will see it with just better-educated decision makers who can make their informed decisions and not assume everything is mystical or things are simply a matter of choice, but that it's actually how part of our preference structures are formed.[105]

He ended by reflecting again on the role of his sexuality studies in liberalizing politics and changing policy:

> Probably gay rights is one of my favorite ones. . . . If you said to me in the '80s or '90s that there was going to be a strong movement and a public awakening and understanding that sexuality is not some choice and folks engaging in it aren't simply amoral, crazy people, or that there's a right or wrong way to do it, I would have said, yeah, no way, the public is never going to buy that. And then of course science comes out and changes the handful of minds. It hit the State Supreme

Court judge in I think it was Hawaii first, who then read the piece and said, well you know, there's some science we're going to take into consideration. And the state justices in Iowa say, well, for scientific reasons, this is the reality. And then the Iowans change their mind, of all places in the Midwest that you see state polls come out. Midwesterners go, well, you know, we really don't believe that gay sex thing, but apparently it's natural, and just hearing the public talk that way, it's kind of a head scratcher. And then of course we are starting to see the states just move in unison to legalizing gay marriage.[106]

Another was more frank about the potential misuses but also suggested that knowing the truth was better than not knowing it:

Maybe there's going to be some marker out there that we haven't found that has a large effect on a social trait. Which seems to be totally impossible to me, but maybe it is possible and we just haven't found it yet. And this marker might align with a certain sex or ethnicity or something else. Even if all those things come true, private industry will be the first to know it. The public will be the second to the academy. Better that the public be aware of it, if they are not aware of it. . . . Wouldn't you rather everybody know about it than only the handful that have control?[107]

Again, social genomics was offered as the solution to society's ills around sexuality.

Sexuality, in sum, is at times portrayed as something genetically in question. Race and gender are not. What this shows is that the political context of how a trait has been debated in the larger culture makes for its presentation as more or less overtly deterministic, and more or less rooted in our DNA. Still, essentialism is an ever-pending presence that affects the characterization of traits in an interlocking way. Increasing essentialism in one domain has already been shown to increase essentialism overall.[108] So race and gender homogeny will create a more biologically essentialist world in ways in which sexuality will also come to be looped.

Determining Essentialism

I like to think on sociologist Ann Morning's work on essentialism around race to understand what's at stake in these sociogenomic ways of determining difference.[109] Her survey of biology textbooks published from 1952 to 2002 has shown that scientists depict race in essentialist and antiessentialist ways.[110] While many, like those shown here, talk about race in rote deterministic ways, many are aware of the dangers of doing so, and so they offer up countervailing biological evidence. Yet among these factions, almost no one talks about the social construction of race, its roots in historical events, and its making and remaking in the past and the present. In nearly all accounts, biological (and increasingly genetic) evidence is the arbiter, the defining factor of race.

This means that essentialism isn't really eradicated by arguments that use DNA findings to argue whether or not race is real. In fact it's the opposite. DNA, with its portrayal of the truth of human being, wins the day. So essentialism can thrive and amplify even as arguments are assailed against it.

When it comes to race, gender, and sexuality, and any other major social axis of human difference, identity is at stake.[111] As sociologist and historian of science Alondra Nelson has argued:

> The special status afforded to DNA as the final arbiter of truth of identity is vividly apparent in the language we use to describe it. . . . Hyperbolic phrases such as "code of codes," "the holy grail," "the blueprint," the human "instruction book," and "the secret of life" suggest a core assumption about the perceived omnipotence of genetics.[112]

DNA evidence isn't innocent. Its invocation calls on an array of assumptions deeply rooted in society, especially but not limited to the cultures where sociogenomic knowledge is being created.

Most importantly, DNA evidence has the power to change social relations and even social structures on the basis of its interpellation of identity. Anthropologist Paul Brodwin reminds us:

> We must therefore ask, how does new genetic knowledge change the ways people claim connection to each other and to larger collectivities? How, in turn, does this process change the resulting webs of obligation and responsibility: personal, legal, moral, and financial? Knowledge of genetic connection alters how we imagine our "significant same": those people who are significantly like me, connected to me, and hence the same as me in some categorical sense.[113]

We mustn't lose sight of the ways that claims about DNA, which are inherently part of an encroaching geneticization of traits, can shape our social life and affect our material circumstances.

CHAPTER 5

THE BREAKTHROUGH

Sociology, they built their own castle for a hundred years. They
built their own terminology and then they've been using this kind of
terminology to bring or show to other folks how everyone is socially
constructed, everything is socially constructed. . . . But now, we cannot
do this anymore. You folks are still talking about your theory on how
people or humans are socially constructed, but the data are there. The
genetic markers are there.

—Sociologist, Chinese University

As far and wide as social genomics studies reach, there is one thing that is common among all researchers. Study leaders are confident that social genomics
is bringing about radical changes in the sciences, both in terms of disrupting
staid notions in natural science and overturning inadequate models in their
own home disciplines in social science. Here I take us deeper into researchers'
worlds in order to explore this unique intellectual spirit of the field. I show
that study leaders see themselves as reformers of mainstream genomics, and
trailblazers toward a more comprehensive, insightful science of the human.
What you'll see is that most social genomics researchers, even the junior professors and trainees who feel that they are unable to carry out their research
to the lengths that they desire in the current disciplinary climate, are awash in
pioneering optimism. They even hold an optimistic sense of oppression and
constriction that feeds into the feeling that they are pioneers of a new future.
This posture aligns with the charisma-driven style of thought by which others
have characterized the present age, and which I have described with regards to
the sociogenomic orientation to the world, which pivots on proactive positive
responses to the surrounding world.

Blazing Trails in Genetic Science

In my conversations with social genomics researchers, I encountered several different kinds of explanations for why people got into the business of gene-environment science. The first thing I learned was that researchers see the genetic sciences as having inept models and they believe they can improve them. For one, they see geneticists continually coming up short in the domain of statistics. Economists were especially vocal about this, as one professor at an Ivy League university known for its biology programs illustrated when he said: "In a lot of ways, actually, the statistics that geneticists use is pretty simplistic from an econometrics point of view."[1] Sociologists and political scientists too said that their understanding of statistics via large-scale longitudinal cohort research and national polling was more sophisticated than that of geneticists. All in all, social genomics researchers described statistics outside of the social sciences as "crude" and bereft of "formal models."

Yet more than statistics, study leaders said that genetic models haven't had enough to say about mechanisms. As the Ivy League economist further argued:

> The usual approach in genetics is to just look for associations, basically run regressions, where you're finding associations between variables. And one of the things that's a little frustrating to an economist about doing that is that you don't get an understanding of what the mechanism is. I mean is this SNP [single nucleotide polymorphism, or "gene variant"] associated with smoking behavior because this person, a person with more of some variant of this gene, enjoys the nicotine more? Or is it that they get more addicted when they try it? Or is it that they have a harder time quitting?[2]

This researcher felt that it was irresponsible to stop short after a genetic association was found. He also raised the issue of feedback loops in behavior and biology. He suggested that genetics could, as one German economist put it, "learn from us as well, especially when it comes to statistics and fiscal methodology,"[3] to "actually model behavior and people's behavioral reactions and how people react to their own genes" even if all the information they had was family history.

Many social genomics researchers feel that they have a better sense of when DNA should come into the picture in gene-environment science. One political scientist at a northeastern American university said: "I usually start with a lot of time spent on the measure, on the social environment, really trying to clean all that out, and basically get my trait or question to a point where DNA is useful. If you start with DNA, it's not very useful."[4] A sociologist at a social genomics research hub called this search for social effects the social genomics cache: "It's a different way to think about gene-environment, to think about a gene that it generates effects, rather than always just looking for the gene for higher or lower IQ."[5]

Another sociologist located at one of Europe's leading research institutions pointed to the social genomics researcher's superior comprehension of environmental factors:

> So what I would hope to see that I haven't seen as much yet from the biology and the genetics side is that there is more realization of what the social sciences can bring to this in terms of our operationalization and understanding social mechanisms, which is very strong. . . . They think, oh we can just measure socioeconomic data—we know how to do that. And you think, no, you actually don't know how the mechanisms work. We have spent years, decades, understanding how this works and operationalizing it, measuring it.[6]

These researchers see their gene-environment model as more fleshed out, taking the environment more seriously. But they also feel that their approach takes the *biological* environment, or the biological system, more seriously.

One research associate who works in an economics department and a hybrid gene-environment research institute in Europe spoke of his experience working in a brain genomics group:

> I don't think they know what the brain is for. Fundamentally, they don't. They understand [it] like this circuit and that circuit and this data, and they have reduced it to a particular level. My favorite example of this is being at a lab group meeting where they were talking about the Resting State Network in the brain, and there are a thousand papers written on the Resting State Network, the Default Mode Network, and studying the brain while it's at rest. (It's a really cool experiment. You just see how people scan your brain image. I mean you can see what's going on there. It's a really simple experimental paradigm. You can actually run thousands of people in it.) They all talked about all these great studies and nobody, *nobody* was sort of saying, okay, what is this for? What does this system do? Why is it that metabolically the brain consumes 20 percent of our calories, this Default Mode Network consumes 20 percent of those calories? So

it's really calorically expensive. It's insanely expensive. Why bother spending so much caloric resource on this network that turns off every time we ask people to do technical tasks?[7]

This researcher argued that it's because of humans thinking according to their social nature—"who agrees with me or disagrees with me, have I offended anybody, what I am going to eat for lunch but who is going to go to lunch with me, all of that kind of stuff." He listed other questions that went unasked like, "Why do we have the epigenetic mechanisms that we do that don't seem to be present in chimpanzees? Why do we have maybe the millions of genes that we have in our gut bacteria?" The neuroscientists in his network weren't asking those critical questions; therefore they were missing the bigger picture of the systemic causes of brain chemistry.

A senior sociologist and administrator of a major cohort study talked about this in terms of the published literature and the limits to which geneticists have sufficiently attended to the real causal factors and their effects in the body: "You look at titles of papers and they say 'DRD2.' And [the SNP] is actually in another gene called ANKK1. . . . But the thing is, it's not clear that ANKK1 is expressed."[8] This sociologist took geneticists to task for making it seem like they knew the gene for specific behaviors, especially without checking whether and in what organs the culprits might be expressed. He referred to "PubMed" as the Wild West—a domain in which any incomplete gene-environment interaction could be published.

As a journal editor of a top-ranking social science journal echoed:

Even if you did find [a genetic variant] that had more predictive power than we typically find, you wouldn't know that it was causal. So for example, you might say, "Look, here's a SNP that's correlated with educational attainment!" But maybe the reason that it generates education attainment is not because— well, who knows what mechanism is at play? Is it that this SNP causes people to be more ambitious and *that* causes them to go further in school? Is it that this SNP causes them to be more intelligent and that causes them to go further in school? Or is it something that is entirely environmentally mediated like, this SNP increases competition so it encourages high achievement in school, meaning getting high grades, but it doesn't have anything to do with educational attainment where grades aren't part of the picture (so only on a kind of competitive ranking-based system do you get a correlation, and once you will remove the competitiveness and rankingness of it, it doesn't affect educational attainment at all)? When you find the SNP that correlates with an outcome, you don't know whether it's a causal SNP or whether it's working through some other factor that happens in this environment to be related to educational outcomes. Or, and here's the real corker, it's just correlated with another SNP on the genome that's actually the causal SNP, and this SNP has nothing to do with the outcome in question.[9]

After explaining the pitfalls of the state of the art of the genomewide association approach, he asserted that social genomics researchers were "really hesitant people" who had a better handle on what was really being found.

Several people I talked to wanted to upend these limitations to benefit behavior genetics and genetic branches of psychology and psychiatry. A political scientist at a West Coast American university said:

> There is a whole revolution now in the social sciences, a move towards experimentation. And in economics also a move towards natural sources of random variation, . . . but behavior geneticists are not in the position, in many cases, to do an experiment. They have to rely on this observational data. So they have to be very, very careful to think about all the different potential confounds.[10]

This researcher extrapolated the need for greater attention to our social nature to experimental design throughout the sciences. A sociologist at a midwestern American university similarly complained about the lack of longitudinal analysis in behavioral studies, but also her positive experience demonstrating the sociologist's perspective:

> We look at events, how they influence later events in life, and I mean this is the core of what's happening, in parent-child interaction, all of these things that we look at in sociology are actually quite important. And we always see things in a social structure and its context, and that so differentiates us from psychology. I have to tell you because you are a sociologist; it's like these basic things that I find myself explaining to people. And then you talk about the life course, we talk about you know events and all these things. How we measure them, how we look at them. And then they understand why you might have some interest and some value.[11]

An economist who runs a major social genomic lab explained how his lab works to simultaneously teach behavior geneticists and genetic epidemiologists how to do it right:

At the moment what is happening is we're basically just learning from their mistakes and from their solutions to these mistakes, and we're trying to spread the word among the social scientists who are very eager to analyze these genetic data since they're out there. They're sort of begging to be analyzed and if you do it the wrong way then you get a proliferation of irresponsible and misleading research results, lots of false positives. . . . We're trying to educate them and tell them how things should be done. I don't know to which extent we're successful, but I hope that eventually we're not going to have as many false positives in the social sciences as they had in the first twenty years of genetics research and medicine.[12]

He concluded that 90 percent of the research done before 2008 was rubbish due to false positives. But he also blamed other fields for misappropriating economic variables like socioeconomic status, which he saw as slapped on as instrumental variables without a complex understanding of their nature.

Finally, researchers expressed concern for leaving behavior genomics in the hands of behavior geneticists because they worried that behavior geneticists had produced so much unethical science. As one education specialist said: "At times I try to justify that working in genetics, given the kind of unsavory history that some of these things have, and I just feel like to leave the field entirely only leaves it open for people that are less careful, less scrupulous."[13] Or as this health policy expert argued:

Genetics are part of medicine now. In my lifetime, they will be part of public health and social policy, too. The question is, how are we going to manage that? I think one of the things that needs to happen is that social scientists need to develop a twenty-first-century understanding of the genome. Right now it's this twentieth-century understanding of the genome which is eugenics and it is genetic determinism, and we know that that's not the right model. We know that there is more information to be had on the genome that can help us understand the kinds of social problems that we want to tackle. The information that's there isn't a blueprint, it's not the whole story. It isn't destiny. There is something we can change. So cultivating that understanding of the social sciences sometimes is more important than any specific empirical project people will undertake.[14]

Finally, a leader of SSGAC research drew my attention to the major misconceptions that behavior genetics has propelled in its concept of heritability and its relation to its history of discrimination:

There are a couple of things that we try to emphasize always, when we talk about our work to outsiders, that are not so terribly obvious at first sight. There's this misunderstanding that if you show that something is partly heritable that this somehow reduces the potential role of free will or the environment. That's absolutely not true. That's just a conceptual misunderstanding. As a matter of fact, these heritability estimates, they don't put an upper bound on the relevance of the environment. The relevance of the environment and the genetic factors do not add up to one hundred. As a matter of fact, heritability can be induced by the environment, and that is incredibly important to understand. So that's number one. Then, the second aspect is that—I mean, some of the sensitivity is really warranted because there has been a horrible history, of course, about using or misusing, half-using science for discriminating people and for all sorts of horrible purposes. This is one area that we really, really, *really* need to watch out for, and scientists also have a responsibility for not only taking this issue seriously but also stepping up and actively doing something to reduce that something like that may happen. Maybe to some extent, the BGA [Behavior Genetics Association] community hasn't taken the stand actively and strongly enough on the past.[15]

This researcher went on to say that he still felt a part of the behavior genetics community, and that he felt the social genomics presence was teaching it better methodological practices. In fact he credited social genomics with moving that field toward molecular methodologies and better use of the data "because we're now in the scene and we're participating in these debates."[16] As with many other of these social study collaborators on IGSS and SSGAC publications, he fashions himself a methodologist with ethical foresight, one who can bring a socially responsible vision to gene-environment science writ large.

Whether attempting to improve some specific area of genetics or biomedicine on whole, social genomics researchers were enthusiastic about contributing better informatics to all disciplines. Talking about big data and the use of Twitter datasets and electronic medical health records, one Canadian economist said:

I program my code by hand. I don't use CAM [computer-aided manufacturing] software. If you look beneath GWAS studies, they're using CAM software like PLINK. And yet there are some things [that are connected] like new genetic effects, like the effects and number of risk alleles. And people generally look at one health outcome, let's say depression, made through a GWAS. But people who are depressed are also more likely to be obese. If you find a gene that's correlated, how do you know the depression is really not just coming through obesity? So using the social sciences, we have these things called "seemingly unrelated regression" or "estimate existence of equations," allowing for those correlations.[17]

She insisted that social science computer models made for better GWAS analysis. Another economist at a midwestern American university argued, "We have the advantage with something like the [Health and Retirement Study] data that is linked to Medicare claims, to actually look at realized measures of healthcare use, not necessarily ones that are self-reported and that could be highly biased."[18] This researcher pointed to the unique big-data resources and protocols that social science has to offer. In fact, one UK-based political scientist described creating totally new methodologies like gene-environment computer simulations:

It sounds insanely complicated, but we know this. We know all of these things from the genetics literature that, once I have written the code, I can just plug in the values. And it's like, I know how this part should work, I know how this part should work, I know how this part should work, but just nobody has ever stuck all of those parts together. And computer simulation allows me to do that. And biologists are doing this all the time; they are doing biological simulations. I am inspired to do this precisely because they are developing simulations in the heart where they've got cardiological models, they've got fluid dynamic models, and they are putting all of these different models of each bit as they figure it out into one framework and seeing how all these bits work together. And I want to do basically the same thing that biologists are doing in the political science context.[19]

Researchers like this one hang their innovation on informatic development, because they are after individual-level differences. And as one economist put it, "At the intersection of the population in genetics, I found the predictive power at the individual level to not be that strong."[20] These researchers see the postgenomic revolution as uniquely information based, evidence based, and big-data driven.

Score One for the Home Team: Toward a Better Social Science

Just as much as social genomics researchers concern themselves with teaching the "other" (the bioscientists who are involved in genetics and genomics), they concern themselves with enriching their home disciplines and the social sciences more broadly. Many researchers see their work as pushing just the right boundaries in both, but they especially care about advances that they see as urgent to the evolution of social science.

A number of researchers described their science as the logical extension of their field's work. In one instance, a European economist took me through the field's biography, starting with its humble beginnings as history and sociology, on to its adoption of math, game theory, and then psychology, stating, "The next step for me? It's so obvious that it's biology."[21] Another European economist who is based in the United States said genomics was the corrective to the inadequate behavioral models of his field:

> Until about thirty years ago there was a wide agreement that the model of decision making that we had was roughly satisfactory. . . . Then there was the period of ten years in which there was great confusion under the sun because people were running (I was one of them), making one model after another, trying to "we incorporate this, we incorporate this, we incorporate this, and here is a new model!" So on Friday, you had the new model. On Monday, there was new evidence, so on Thursday you'd have to make a new model. . . . So I think many people realized that we needed to look under the hood to understand how the engine works rather than cooking up more.[22]

But some researchers described an unfortunate lag in their field's adoption of this *inevitable shift.* A senior political analyst at a southwestern university said political science had been at least "a decade behind" as far back as the 1990s:

> We were kind of applying 1950s survey research from psychology, but it's like 1990 and psychology's moved along and has learned a great deal since then. . . . Economics had moved into political science in the form of rational choice approach and gain theoretic approaches (again, our characterization of economics as we were applying it in political science really didn't take into account the whole experimental or behavioral economics revolution that was taking place in economics). So it's almost exactly at the time that kind of cutting-edge economists like [Daniel] Kahneman were discovering that there were real problems with a kind of strict rational choice model of human behavior, because humans just didn't behave that way that political science was still busily exploring. All the exciting things you could do in political science with that very same kind of thin rational model? That was really no longer cutting-edge economics. And so . . . I had some interests I explored with a group in Australia, doing some work that involved some actual genetics work we were interested in, and some cross-variation of vasopressor receptors. But we were very quickly disabused of the

notion that that was either easy to measure or could be done entirely accurately or really would get at a full picture of what it was we were hoping to find on the independent variable side. Again, that's led to most of what we do now.[23]

Though this scientist felt "ahead of the curve" since getting involved with geno-politics, he saw the search for what is under the hood still arriving ten years late.

At base, social genomics researchers described biology as the birthright of social science. That may sound contradictory given that "the social" is typi-cally connotative of "the environment," but it is because they truly believe that biology is in the environment just as the environment is in our biology. These remarks by the education specialist above capture the common sentiment:

I just feel like if you really care about certain social things, like if you care about how and when people get married, biology is such a fundamental part of that. I just don't know how you could ignore it even if your interest is not in the biol-ogy of it. I get that some people are interested in the social processes that put people in connection with each other. But I think at the end of the day, you can't get rid of the biology. That stuff is just fundamental.[24]

A political scientist at a southwestern American university argued:

There isn't a parsimonious way to build humans developmentally without tak-ing advantage of the fact that you have both genetic code and then you have an environmental envelop that they'll develop in. You see that in all kinds of areas of developmental genetics, and it's certainly true for areas of social beliefs and behaviors as well. In that sense, it's true that the notion of a nature versus nurture is a silly distinction, because there just isn't a concept of developmental genetics in which the environment is not a crucial interactive piece of the mold-ing or building process. At the same time, that doesn't excuse how in the social sciences most people initially thought genetics have no role. There's now some move toward the view that there is a kind of a gene-environment interaction, but that's mostly used as a way of dismissing the importance of genes.[25]

This political scientist is part of the ranks of researchers who are concerned that the social sciences are still only paying lip service to the DNA revolution in science.

What's more, these scientists believe that socialization has been overstated in social science. As the sociologist quoted in this chapter's epigraph put it: "We're talking about the importance of parenting on children's behavior. Well, parenting is important but how about the genes? People have some kind of particular marker which is associated with a high propensity of delinquency. Even good parenting might not change it."[26] A doctoral student focused on health economics tempered this statement, asserting the differential susceptibility theory, again the theory that maintains that our genes make us susceptible to environmental influences in varying ways:

> You look at lower-income kids and genes don't seem to matter, or hardly at all . . . when kids are really poor and the environment is really challenging. So they are breathing lots of particulate matter and their parents are smoking lots of cigarettes and there's lots of cockroach dander in their homes and they haven't seen a primary care physician maybe since they were born. In that type of circumstance where environments are really bad, genes don't matter that much, because the asthma rate is forty percent. Just everybody has asthma. And then as you move into better things, people with certain genomes might respond to those things whereas others might not. And then when you get into a situation where all the kids are kind of having their basic needs met, well the only thing sort of left to differentiate them is their genetic profile. So the genes are all that's left at some point. Asthma prevalence rates are much lower for everybody, irrespective of their genetic background, but the remaining differences across kids might be explained more by genes.[27]

As a leader at the SSGAC summed up, understanding genetics was part and parcel of understanding "how genetics could contribute to economics" and eventually "realizing some of the potential payoffs for the discipline."[28]

Better empirics in one's home discipline is indeed the resounding mantra of the field. As the cohort study leader mentioned earlier argued:

> You know, there are some people who think of it in really dichotomous and zero-sum terms and that the genetics people are taking away territory from what sociology is about. But then you're supposed to sort of move beyond that to interactions, and the ways that those interpenetrate one another is so fundamental and thorough that there's a level beyond the way that we're able to talk about it . . . as a theoretical project. But then the empirics of that are going to be in the relationship between psychological traits and attainments in life. And [our cohort], . . . they've lived their whole lives. So you can see how their whole lives have turned out.[29]

This principal investigator moved from the challenges to sociology and beyond to end up with its strengths—its mastery of complex biosocial data—because he wanted the discipline to live to its fullest potential.

Similarly, the West Coast-based political scientist quoted above simultaneously cheered and jeered his field for its statistical prowess:

> The statistical analysis itself is like almost exactly what we do as a social scientist. What's different is that there are a number of considerations that you have to take into account. Population, that's the number one thing, right? Realistically, we probably have a population stratification problem in all of social science. This is one of the beautiful things about being able to reach out to another discipline and do this sort of crossbreeding and the systematic learning about the genetics and learning the way that they do their work. I think it's made our pure social science work better as well.[30]

This scientist now thinks of himself as a "data scientist." He teaches political scientists to avoid doing regression on regression without controlling for multiple testing, and to see genomic tools as "generic," as simple data points in matrices of datasets genetic *and* social.

Many social genomics researchers also say they want gene-environment approaches to help eradicate what they see as unscientific leanings in social science. A graduate student at a "Research-1" (top-level) American university said:

> I think the most important thing that could sort of spill over . . . is if the social sciences adopted a more rigorous scientific method like the sciences do, and that is writing up analysis plans beforehand, submitting them, not data mining (which is what I feel like half of psychology is these days) when they have this severe publication bias geared around null hypothesis testing, and [determining] whether the P value's under 0.05. You probably read recently, there was one psych journal that basically banned that sort of "no hypothesis testing," trying to mitigate exactly this problem.[31]

This student illustrated the fast and loose picture social genomics researchers see when they look at their home disciplines. Researchers also describe theirs as a move toward the more robust collaborative science that leads the day. As expressed by an SSGAC leader who has been instrumental in exemplifying the modus operandi of social genomics research, getting funding for large-scale international collaborations between scientists of all ilk: "There was a time when you could just make scientific progress by being incredibly bright and sitting in a room with a paper and pencil. These days I think that's super hard. And a lot of the advances in science come from collecting good data."[32] He and his collaborators hope to introduce new rigor, and eventually new publication and funding standards.

Ultimately, these researchers want to make sure that their disciplines will not be left out of the DNA revolution. As a political scientist and early pioneer quipped: "My concern is that political science will really get left behind as a discipline. And it's fine now. But in ten years, it's going to be obviously left behind in funding. They really are not catching the wave." Given the dwindling social science funding portfolio, many social genomics researchers have come to see gene-environment science as a join-or-die situation.

Sticking to One's Guns and Following One's Heart

You have now heard some of the intellectual motivations of pioneering sociogenomic scientists. But researchers also have other, less overtly rational motivations for carrying out social genomics work. I asked study leaders to tell me about how they first got interested in genetics as a more holistic way to understanding what is in it for them. What I learned was that scientists have been spurred on by a nagging hunch that biology is fundamentally important to all science, and a strong sense that iconoclastic science is the best way to do science.

One economist excitedly described coming up with the idea for genoeconomics with his advisors:

> In 2001, I was at a conference with my advisors. At the time, I was doing a master's degree in England and I was about to start a PhD program at Harvard, and two professors who I knew from my undergraduate days at Harvard, who would soon be my advisors at Harvard, were at this conference with me. And at the time neuroeconomics was just beginning, so people were curious about it and interested. At this conference there were some presentations about neuroeconomics, and so it was the hot new thing. And me and my two advisors took a walk in the evening after the first day of the conference and we were talking about neuroeconomics. And then someone raised the question of what would be the next big thing after neuroeconomics, and we thought genoeconomics! And so we decided at that point that we would start working on it.[33]

He expressed the common sentiment that science is about pushing the boundaries of what everyone was just getting comfortable with, and that it takes bold thinkers to move science forward.

A sociologist who began with genetic studies of delinquency similarly said:

We learned a lot on our own. We read widely. We e-mailed people and talked with professors from other departments at [a Research-1 American university] or at other universities and on the phone a lot. We basically learned everything, self-taught. It was very challenging in the beginning because when we first started getting into this, at least from the field of criminology, nobody else was doing it. It was just the two of us really doing this line of work. Now, there's a wider range of people that do it. It's still a really, really small group of individuals, but we're more interconnected with other fields of study, have worked with people. Some of my former students have gone to workshops, certain of them taking classes from other departments have actually done some genotyping and whatnot, so to speak. It's moved from sort of a grassroots effort to something that's a little more institutionalized. Certainly, we had to learn all the methods on our own and whatnot. It was quite challenging. It still remains a challenge because it's like we're straddling multiple disciplines, and so there's not a clearly carved-out genetic way of presenting results for criminology; the way that we present results, say, for a psychology journal would be different than the way we do for criminology, different for genetics and so on and so forth. There's still the challenge of them. But absolutely, we had to learn on our own.[34]

In all cases, this sociologist has bet on his hunches and gambled with the resources he's had to do what was in his heart.

In fact, many study leaders emphasize the processual nature of innovation, invoking the adage of always following one's gut. One political scientist, for example, described his cautious excitement at hearing about gene-environment political research for the first time at a conference, but pursuing his instinct as he worked to make sense of it:

There were a few scholars who wrote about biopolitics in the 1970s and 1980s. The problem there is that they, in those articles, asserted that biology played a role in politics, but they didn't test it empirically, and so they ended up creating their own journal, *Politics and the Life Sciences*. But it ended up being a relatively noninfluential journal and a noninfluential group. . . . So I saw this presentation, I believed it, then I went away and started thinking about it. And I didn't really understand the method, because I just had never thought about the twin study method before. It's not the most intuitive thing in the world, and I disbelieved it. Then I went back and read the paper and then I believed it again. And then I disbelieved it. But I kept sort of coming back to it. And finally, like this is something that clicked, and I thought okay, this has *got* to be true.[35]

One economist based in Canada called social genomics "a frontier research" less interested in quick answers that generate "ultimate truth about life" than "to spur additional inquiries, to dig deeper, to go broader," following one's nose to more innovation and deeper questions. He said:

> I think about it as a process, as a much more long drawn-out experimental process where we are only just beginning to tap into the very tip of an iceberg and we have no idea how deep it really is. It might not be very deep at all or it might be any realm of iceberg.[36]

Indeed, pushing forth with training despite disbelief from others is a big part of social genomics researchers' stories, evincing the frontiersman disposition permeating the field. As one genopolitics pioneer explained it:

> I am optimistic that because of the influences genopolitics has had on increasing awareness of an interest in all of these different kinds of algorithms from genes all the way up to just behavior, we observe everyday that this is here to stay and that it's really opening up the next generation of the normalization of political science as a science, which is "How do we integrate what we're doing into what they're doing in psychology, what they're doing in biology, what they're doing in biochemistry?"[37]

Graduate students likewise declare feeling empowered to carry on the baton. One grad student at a private social science university with no social genomics program said:

> I feel like this is a great way for me to do something that is unique and different but really blends my interests in econometrics with also wanting to take a more life-course view on aging. . . . It's like why would I not bring the human genome out of the air and into term? That's how I think of it as an econometrician of course, which is again a very narrow way of thinking about it. But it seems like, oh gosh, the data is there, how could I not use it? How could I not incorporate it when it lends dimension to the work?[38]

Another grad student at an Ivy League program that also has no social genomics program talked at length about changing the public understanding of science, showing the confidence that students and educators alike take in doing what they call "science at the cutting edge."[39] The common thread among study leaders is the notion that science is an adventure best pursued by the brave and the sure.

Disposition's Impact on Discipline

Social genomics is not the first (nor will it be the last) field to launch with a risk-taking fervor. In *Misbehaving Science*,[40] Aaron Panofsky characterizes behavior genetics as a field brimming with mad cowboys set on pursuing unpopular research agendas despite pushback from the very disciplines from which they arose. Panofsky argues that as a result, the field of behavior genetics has formed into an archipelago, a chain of disconnected research enterprises holed up in long-standing disciplines like psychology and genetics. Due to constant controversy, researchers have bunkered down and narrowed their research focus to the topic of heritability. They have sought public notoriety in place of scientific notoriety since the wider science community has been hostile to the field.

I was interested in how social genomics was faring in this regard. What have been the consequences of pioneering hybrid science? Have social genomics pioneers felt supported in their endeavors? Have they been working within the framework of the scientific mainstream? Or, more like behavior geneticists, are they lone wolves who stick to their own individual careers?

Study leaders after all have described a number of scenarios in which they faced great pushbacks. Some talk about facing pushback in terms of a general suspicion about genetics, and potential nonmodifiable or anti–free will aspects, as one education researcher illustrated when he had to squash publishing in a big science venue to avoid tipping off his colleagues who would "think it's unsavory."[41] Others talk about getting lambasted in reviews, like this one political scientist who was called Hitler:

> I got one of these reviews this morning, and that's why it pisses me off so much. One of them is like, "If you're doing genetics, it means you're Hitler. It means that you're doing this incredibly evil thing." Actually, the people who say that have no understanding whatsoever of the incredible strictures that are involved in using this data. The anonymity, the transparency, I mean, way more than any of them ever have to do with the shit that they're doing. I mean, if they had to put up with their little pieces of experimental data, even one-tenth of the constraint, they would flip out.[42]

Still others complain about the lack of science education in their home discipline constituencies. As one Middle Eastern economist working in the United States put it, "People are not sufficiently familiar with evolutionary thinking."[43] The political scientist who was called Hitler agreed:

These are people who I doubt passed high school biology. They really are part of the endemic American problem of having a little bit of fear of science. It's this complete marriage of fascination and fear. There's this fascination like, "Well, if you can make me lose weight by taking a pill, that's awesome. But God forbid that you actually use genetics to study politics."[44]

One of her collaborators put it similarly:

The majority of people with a PhD in sociology or political science have never taken a science class, not once. (That blows people's minds in the sciences: the idea that you can have a doctorate degree with never having taken a class in biology.) And so, when you say genetics to people, that means there's a gene that determines this—that's what it means to them. And no matter what you write, that's still what it means, and what they respond to. This is a portion of people in the field, just out of total ignorance in the worst way. It's not even self-imposed ignorance. It's just a state of it of being able to get a degree, a doctoral degree, without some science class or statistics class in many cases.[45]

He also complained of the name-calling in their field:

That leads to just an instant belief that you're some evil person or that you're going to try to find the gene for liberalism and get rid of it. And so, there's that group where you have a lot of hostility and frustration no matter what you say or how you educate or what you write; they will not change their mind, they're very path dependent.[46]

He continued by explaining how the name calling trickled into publication and grant reviews, and tenure and promotion letters, and then expressed the prevailing perspective that the journals, funding, and tenure review processes were "a century behind" the times.

But overall I found that despite numerous setbacks, social genomics researchers have managed to rise up through standard mainstream channels that have allowed them to build their career profiles *and* the field even as they have taken great risks. First of all, researchers described seizing upon their institution's interdisciplinary opportunities. As one sociologist at a midwestern American social institute said:

I've gotten money to do different things from every level at the university. Don't get me wrong, it's harder to get money from the university, the broader university, but they've given us money, and they're giving us money to do other things and to get together in different ways. Yeah, so it's pretty supportive. I'm proud because [my university] tends to be a very interdisciplinary place. It's like almost everyone seems like they're in one or two departments. Like two or more departments. . . . Sociology is not. I'm not in sociology anymore. I got my PhD in sociology [here], oddly enough. Now, I'm back there, but I'm in a different department. There are some sociologists that are interested in it, but those are all demographers at the [demography institute], and so that's who I interact with.[47]

Another sociologist described taking advantage of support from her university, which is pushing interdisciplinary work: "They're like, 'Yeah, we want a sociologist on this grant application.' But within the department, people are either fine with it and not interested or very against it or very defensive about it."[48] She believed that any disciplinary hostility was due to "the old guard, the established people who probably fought really hard to get sociology widely recognized," who she understood founded the field as a counter to biology.

Social genomics researchers have also followed a strategy of targeting the top disciplinary journals in the natural and social sciences and publishing in both. As one American sociologist working in the United Kingdom summarized:

Journals are not to a large extent transdisciplinary, so you have to pick your camp. You have to publish and publish high in sociology or genetics or science journals, of course the top journals, but I think in the end you are still forced to publish in a disciplinary manner, by the department you are in; you are ranked and evaluated in terms of publishing in a discipline. So there is a whole structure there that even if you have the change with the students, you need these hybrid students, [but] they come into a rather disciplinary-specific world.[49]

Instead of looking to create alternative publication venues or hitting the streets with their findings, they have worked informal connections and formally appealed to the top tier in a range of disciplines, making their science heterodox and orthodox at the same time.

Another thing researchers have done is work invisibility to their advantage. Many social genomics researchers said that their home disciplines were ignorant about much of what they were up to. As one health policy specialist put it, "Many mainstream economists just are pretty indifferent about it and think of it as pretty niche."[50] He later said that his colleagues in his department also mostly "don't care."[51] Researchers also talked about capitalizing on the ignorance in their jobs. For example, senior scientists echoed the sentiment expressed by one Scandinavian economist that being "a full professor tenured" meant that one could "freely pursue [one's] research interests."[52] Or as one Canadian health economist said, "As long as I bring in money, who cares what you do?"[53]

It did bother junior scientists that they have had to develop a two-track career. Many talked about tenure anxieties, like this UK-based economist who works in a politics school:

> In the UK, every five years is a research evaluation framework and every depart-
> ment has to submit for each faculty one or a few papers that are representative
> in the best work. And when my case came up, I said, "Sure, we should put in the
> *PNAS* paper, the *Science* paper." They're like, "No, why don't we put in your elec-
> toral studies paper or your political psychology paper." I'm like, "Do you guys
> realize the impact factor?" "Yes, but the committee will not know," et cetera.[54]

This junior scientist voiced the complaint that home departments don't compre-
hend how much of a splash his social genomics work is making in mainstream
bioscience, and so they are not rewarding for it in promotion. But like others,
he pointed to the successful tenure cases of a range of colleagues who took the
two-track career tack and only grew their influence in both tracks as a result.

Finally, social genomics researchers have followed the mainstream path in
their teaching. Many have held back on teaching social genomics, or simply
sought opportunities outside their home departments. An economist at the
forefront of SSGAC spoke about this:

> It's not clear to me that at present it makes sense to educate students. I am not
> sure it's fair to the students to expose them to this kind of stuff just because it's
> not a well-defined field yet. It's not clear what kind of jobs you could get. So in
> terms of teaching, I really keep it at a minimum. In terms of my overall research
> portfolio, I would say that at present I spend roughly half of my time on this
> work and then I have some conventional economics research on the side. It's a
> way to buy insurance just in case the genetics doesn't pan out.[55]

As another SSGAC leader said of his teaching:

> When graduate students ask me whether they should work in genoeconomics,
> one of the things that I make sure to tell them is that it's a risky thing to do and
> it's unlikely that they'll get a job in an economics department if they focus in
> that area. . . . I think we may be at kind of a turning point now where there will

begin to be professional rewards in economics for doing this work, but there haven't been until now.[56]

Still, both of these researchers have launched teaching programs outside their home institutions at European research institutes and at Ivy League schools far afield.

Indeed, finding a supportive university has been the last stop for social genomics researchers. For many, this has meant changing disciplines. One American researcher in Europe described finding a more supportive institution to work at:

> There was extreme resistance where I was at. People didn't feel that it was sociology. I had people yelling at me, literally saying that it was too empirical. When I presented my work to our department, somebody walked out and said this isn't sociology. . . . [At my current university], people I think are a bit more used to interdisciplinary research. They knew what they were getting when I came in too. But in where I was it was an internal department that hadn't had much movement in or out.[57]

Many researchers like this one have moved to other countries as well. Still others have crafted themselves as hybrid or monodisciplinary. As one Dutch doctoral student on the market told me:

> I'm now in my second year of my PhD, so in little more than two years my PhD is finished if all goes well. I don't expect the signs of economics to have changed a tremendous lot in those two years. So given that prior belief on my part, I think it might be difficult if I want to really get employed in a classical branch of economics. But that's not a sort of prerequisite for me; I'm quite fine for instance to drift a little bit more in the direction of say quantitative genetics and so on. . . . I think in the long run these fields have to come closer together, so maybe then an opportunity will arise to really become sort of a full-time hybrid.[58]

This researcher, like so many others, is open to becoming a card-carrying geneticist if need be.

How Outlook Shapes Success and Secures the Paradigm

It's important to know that the majority of study leaders express confidence that the hostile facets of the scientific landscape have subsided quite a bit. A sociologist of crime said it best:

When we first started doing this, when I first started presenting, we would have
people that would basically heckle us in the audience, tell us that we should be
ashamed of ourselves for doing this type of research, this has no place in criminology. I've had people tell me that they believe that genetic research matters but
shame on us for doing it. It should be swept under the rug and nobody should
study it.[59]

He averred that things are better now that the field is growing visible and getting the big grants. A recent hire at a West Coast university similarly said that
his home discipline of political science has been coming to appreciate the
science:

I think there's a small group of people who really object to it on various grounds.
There's a bigger group but still not a majority of people who are really excited
by and who have embraced it. And then I think there's a lot of people who are
receptive but don't know what to do with it and so they hang back. Part of that
is if you look at political science there's a wide block of this discipline that can
understandably not see why this would matter to them.[60]

At the same time, he talked about the "silver linings," pointing to the visionary
support of social genomics by the editors of the field's top two journals.

Economists in general felt that their field has opened up to the idea of social
genomics. One said:

Among the social sciences, I think economics is the least political, or at least
the least politically correct. And there is very little concern about saying things
that would get you skewered in the press. So there has never been a taboo in
economics against saying that things are genetic.[61]

Another agreed: "It's not at all taboo in economics, the way my understanding it is a little taboo in sociology, to say, 'This might be largely due to genetic
transmission.' Economists are very comfortable with that idea, that possibility,
at least."[62] Yet another economist boasted about methodological openness:

I think economists love new methods. Economists have no trouble at all believ-
ing that there are genetic pathways that influence our outcomes in life perhaps
through complicated G by E interactions or perhaps through simple additive
genetic effects. Everything is a possibility to an economist, and we have generally very few preconceptions and we're very empirically driven. It's funny the
outside world may see us as kind of ideological. But most economists love to see

new data, love to see new methods. And when you have credible data and you're using appropriate methods, they're very curious to see what emerges.[63]

Where economists see pushback is in other social science disciplines. Stating that economics has had no "taboo" around social genomics, this researcher then said: "Now I can't say the same in terms of my interactions with other social partners from the noneconomic sciences . . . from people coming from physical anthropology, cognitive anthropology, evolutionary biology, and to some extent even social cultural anthropology."[64] I also heard members of such external disciplines referred to as "luddites that are there to try to destroy."[65]

Economists were especially pleased with the way they have successfully published in the leading journals. Many described gaining respect from hesitant collaborators who used to think their project was, as one put it, "bollocks."[66] Others said publishing in the field's top journals bespoke a professional virtue on the part of their home discipline. As one economist summed it up, "Professionally, genes have been pretty kind to me."[67]

Sociologists and political scientists, who are slightly more conservative in their celebration of their home disciplines' openness to social genomics, still equally extolled the virtues of the gatekeepers of their field, touting their publications in the top journals and the overall positive reception from curious elites. As one sociologist exclaimed in 2014:

> I published in [a top sociological journal] in 2008. That's different from [the other top sociological journal paper] in 2008. I have another paper coming out in [the second of these journals]. [Then there's] the ASA conference. Paula England is the president. She's going to organize a special session on biology and sociology. I'm going to be invited.[68]

A political scientist who leads a social genomics lab said, "It's easier to publish outside of political science than inside," but noted that his group has published in the top three journals in the discipline and gotten major grants from the NSF political directorate, from influential foundations, and from the Department of Defense.[69]

Sociologists and political scientists involved in demography additionally cite the positive reception of demography's professional elite. As noted in Chapter 2, the journal of Biodemography and the Population Association of America has sponsored all IGSS events and publications. As one junior demographer described it: "You just have that sort of great synthesis where everybody is really sharing ideas and asking interesting questions and supporting junior people, and really pretty open about sharing things."[70] She described demography as seen as a model interdiscipline that is just institutionalized enough yet flexible enough to host researchers innovating in a wide array of methodologies and theories.

Members of the field also give positive appraisals of the universities they work in, no matter how contentious they see their work in the context of discipline. People have felt especially well supported in the American Ivy League. As one senior economist at Harvard argued:

> Harvard is a place that's very, very open-minded, in at least economics. We have a very diverse department and everything goes. To some economists, they like drawing bright lines and saying, "You're in or you're out." That's not the case here. We have an expansive view of economics, and I think that's common actually, but I think economics at Harvard is particularly expansive. I've received nothing but encouragement from my colleagues.[71]

Another senior scholar at Brown said:

> I have a really hard time living in Providence, but I really feel like my colleagues think that what I'm doing is interesting even when they think it's odd. I mean, I think that they definitely have moments where it would be easier for them to understand if I was doing just straight-up international relations and security studies and so on. But I think that they find it interesting and valuable that I do interdisciplinary research and I feel very supported by that.[72]

Postdocs and grad students also exclaimed open-ended support. One Princeton student characterized the atmosphere in the sociology department as full of curiosity, even when colleagues were fearful or confused by social genomics.[73] A student at Harvard also said "curiosity" was the best way to describe the atmosphere in his department. He told of how his department allowed a couple of social genomics alums to buy out his teaching-assistant time so he could pursue research in the area.[74]

Pioneers also feel duly supported by the British network of universities that includes Oxford, Cambridge, the London School of Economics, and the London Colleges. Researchers at these schools, many of whom are not British nationals, say they have had access to large sums of public funding and a great deal of engagement from the popular British media.

Indeed, European universities have led the way in providing and establishing truly transdisciplinary research environments for social genomics. As such, they have also been cited as leaders in support of the field. As one Dutch economist explained it:

> My department gave me all the freedom to do it, and they supported with the financing for PhD students. They like the high-risk effect and they like the cooperation with the School of Medicine. We even got a million of the inside money to support this initiative, which is exceptional. . . . All the sharks and the competition and whatever, no. The deans with the two faculties and the provost, they love us.[75]

A student at the same university averred that the university supported him in any gene-environment research he could conceive of:

> If I have to give you a percentage, then it's probably around eighty percent of my time that I spend on this genetics things. . . . I'm officially an assistant professor at [my doctoral school], but on a fixed-term contract. Some people call me a postdoc. But yeah, the role doesn't formally exist at the [school].[76]

And European administrators of long-standing cohort studies said that the environment for their research has become more supportive in the new era of gene-environment science.

Even the array of Research-1 schools that have been excited about and open to research, but have provided a modicum less of the basic resources to get innovation going, have been touted as earnestly supportive. One junior political scientist at a private American university talked about how his school had been bold enough to hire social genomics pioneers despite criticism from the wider discipline:

> Some people really hate the idea, and I'm always surprised that they'll just completely shut it down like there's no possible way that it could happen. And then I'm equally surprised by how willing people are to at least entertain the notion that it could be the case. . . . The fact that I was hired here is a reflection of the department and the university willing to give this a shot. And I think there are few departments that are willing to take a chance like this one does on this new literature, on this new kind of like research agenda.[77]

Another junior researcher at this institution said he felt supported by his department, which he believed has always been cutting-edge with respect to gene-environment ideas. What's more, he felt respected by the biologists at their institution.[78] Still others have been happy to find that their private institutions have been willing to count their extradisciplinary publications toward promotion. They've experienced a great "flexibility" around what methods they can deploy in their work, and what they can get internal funding for.[79]

Supportiveness has come in different guises in under-resourced schools. As one sociologist at a public Research-1 university put it: "My sense is that as long as I am publishing in good journals, they don't really care what those journals are. For someone like me, that's really important."[80] Students in public Research-1 schools also applauded their programs' support for their gene-environment forays, even though they've had to seek out support from more senior researchers at other universities.[81]

Finally, scientists at new universities garner extreme support of their social genomics research. Whether private or public, researchers at these institutions say they have carte blanche to study what they desire. As one political scientist at a new West Coast university said:

> It's a brand-new university and there's a very strong inclination for interdisciplinary work. So I'm not sure that they particularly have ideas about genetics, but they like that I do interdisciplinary work. And my colleagues, number one, I think they think I do a lot more twins research than I do—a lot more genetics research—just because of stuff I have done. But I think they're very supportive in the sense that they—I did my job talk on personalities but I reference some of these things in relation to the heritability of personality, and I think that they were excited about hiring somebody who has at least had a foot in this kind of biology and politics world, which is sort of—and at least at the time I was hired—obvious heresy a little bit.[82]

Another researcher at this university also described publication placement as the only thing he has to worry about.[83] Finally, a Chinese sociologist working in a new university in one of the special administrative sections of the People's Republic of China said his university has been so into anything new that they "just don't care" like Western universities might.[84] His university is excited about his exploratory research into cognitive function and intelligence. This researcher voiced the prevailing sentiment that social genomics is on its way to something bigger than the sciences has ever seen before.

How does all this bear on the development of the field and sociogenomic science more generally? We know that support in and of itself doesn't mean that a field will be successful. But optimism certainly contributes to the will to succeed. This new field cheer that has descended over social genomics only ensures more attempts at advancing science in the sociogenomic way. In a way, we can look at it as a kind of "collective effervescence," à la sociological descriptor of mob mentality. There might not be a mob here, but there is definitely a network of interested and excited individuals whose enthusiasm is greater than the sum of its parts. And as the next chapter will show, this effervescence is contagious. It spills out to funders and cohort leaders and it keeps sociogenomic science at the top of many institutions' priorities lists.

So unlike Panofsky's behavior geneticists, here controversial emerging science doesn't equate with a need to bunker down or slip around the science community to directly plead with the public for recognition. Social genomics researchers instead mainstream their way to notoriety. They straddle their way to transdisciplinarity. Therefore, in addition to reconceiving autonomization as about turf wars for a specific object of expertise, we have to reimagine it in this postgenomic era as ecumenical, capacious, and optimistic at all costs.

The end result is the bolstering of a certain sociogenomic paradigm wherein justice, activism, disciplinary boundary crossing, and genetics are melded as one and elevated to the status of a postgenomic ethos. This is indicative of a paradigm that we will see circulating throughout the furthest reaches of society in the following chapters.

CHAPTER 6

A BIGGER, BETTER SCIENCE

In my depiction of the brick and mortar of social genomics, I've laid bare many of the support networks and community ties that make up the burgeoning field. I've also shown many of the motivations behind the science, including its interest in improving empirics and population representation, and its ethos of pioneering optimism. But who really publishes, funds, reports, and otherwise supports the science? And who has publicly pushed back against it? How sociogenomic is the context in which it is emerging?

In my visits to social genomics conferences and study meetings, I've seen the intimate connection certain editors, funding agents, and scientific organizations have had with social genomics. Yet as we've just seen, interviews and private conversations have also revealed challenges scientists have faced in getting taken seriously, seeking employment and hiring others from within their home fields, and conducting other basic scientific activities without departmental support. Here I zoom out from the tight-knit group of members to show the ways gatekeepers in the science community feel about the science. This coverage will show that the field has mostly been met with interest and enthusiasm from the wider science community and open-ended support from the research funding agencies around the world that determine which emerging sciences succeed and fail.

Passion from Publishers

I sat in my seat at the back of the room, checking my e-mail as the last speaker of the morning shimmied offstage and the next speaker walked on. Suddenly, one of the conference organizers called on everyone to listen up. A visitor had an announcement. It was the editor of a well-known policy journal who was calling on attendees to publish in the journal. He made it patently clear that the journal was intent on collecting all the reports from the conference that could be formulated into articles in the coming year. He said that the editors were very interested even if study data weren't "final." He further assured attendees that the journal would find fair and appropriate reviewers for wherever researchers were at in the process.[1] This was but one of the first times I would witness journals openly petitioning social genomics researchers to publish with them.

Indeed, a survey of the places that social genomics features in shows that the research has found enthusiastic support from an array of journals, whether biomedical, political, general natural science, or social. Though slightly more research has been published in social science journals, most research teams have been wooed by a diverse array of venues.

Several special issues of journals have been produced by the social genomics community, and looking at their hosts can tell us a lot about this characteristic range. In 2011, *Biodemography and Social Biology* hosted the field's first special issue.[2] In 2013, the *American Journal of Public Health* also hosted a special issue titled, "Society, Genetics and Health."[3] In 2015, the *Journal of Policy and Management* hosted an issue on "Policy Analysis and Genetics."[4] These issues were the product of meetings in which social genomics study leaders presented exploratory work to each other and built new partnerships. Editors have since visited and revisited meetings offering new publishing opportunities and sealing relationships with particular organizations. In my conversations with them, I've only heard exceeding praise for the field and earnest commitments to continue to feature its work.

But what makes social genomics stand apart from other sectors of science is not so much its ability to draw editors to its flame, but its reception in the top-ranking journals of the natural and social sciences—journals that often don't entertain even the most elite scholars. A case in point is the field's most high-profile article on educational attainment, which was published by *Science* in 2013.[5] Its follow-up, dubbed "Educational Attainment 2.0," was published by *Nature* in 2016.[6] Yet even prior to these publications, social science journals like the *American Journal of Sociology*, *American Sociological Review*, *Journal of Politics*, and *Journal of Economic Perspectives* all hosted social genomics publications. In fact, the first publications from this field were featured in the *Journal of Politics*, *American Journal of Sociology*, and *Science*.[7]

Most social genomics researchers say they've had an easy time getting their papers published in high-profile venues from the get-go. Of course, early on, social genomics research teams were asked to bring out the medical or health relevance of their data. When the SSGAC was looking for a home for its educational attainment paper, it first went to *Nature*, whose editors told it that the effect sizes it was reporting were too small, and that there wasn't enough biology in the paper. The SSGAC had to bring out the cognitive and neuronal function implications.[8] But as one researcher who was early to the punch in publishing articles told me, in the mid-2000s you could get a lot of things published and then you'd get media attention that would feed back into your notoriety (as he put it, you'd get much more play than say for a "comparable kind of international relation study").[9]

Now that the work is known, getting published is different. Researchers say it has gotten a bit harder because there is more scrutiny from reviewers. Some have started to hold out replication as a condition for publication. As one researcher said, "No one wants to hear that it didn't work out or that it was really small predictive power or something like that, when that's actually a finding in and of itself."[10] Many studies have been unable to get published despite what research teams see as keen due diligence on their part.

Notoriety has also been a double-edged sword for another reason. As another political scientist who has published his team's work in leading science and social science venues pointed out, "The general science journal is only single-blind, so you see the name of the person on the paper and it's the very first thing that [reviewers] see."[11] He expressed the common sentiment that reviews have gotten more polarized now that researchers have some name recognition for this science.

Some social genomics researchers complain of a methodological divide that has made it harder for certain research teams to get their work accepted now that most people know the range of how the science operates. They say that the new cohort studies have displaced other study designs, such as candidate gene studies. Studies publishing on secondary data that have five to ten thousand participants are no match for the one hundred thousand-plus genomewide association studies out there.

But overall researchers have found that journal editors have become more open to giving social genomics a chance as more studies have come out. One sociologist chalked the opening up in his field to editors seeing that social genomics isn't eugenics (which he reminded me was a serious threat for many in the field not so long ago).[12] Another whose paper ultimately got published in the top journal in his field said that reviewers at the journal were initially nitpicky down to the level of citations (one stating, "Some of these citations are really old and so this shouldn't be published"), and editors were sympathetic because they were so worried about the sloppiness of the kind of work that went into eugenics tomes like *The Bell Curve*.[13] Still another who also eventually was able to publish in the top journal in his field said, "There's a little bit more acceptance particularly of the younger folks who are coming up, who have sort of grown up in the age of the genome and the decade of the brain and understand the importance of biological processes to behavioral phenotypes."[14]

Interestingly, most find it easier to publish in the big-name generalist journals in their field. Researchers say turf wars in the subfields of their discipline have led to specialty journals booting them out. As one political scientist reported, "Two journal editors in particular were incredibly forward thinking, John Geer and Rick Wilson [the editors of the *Journal of Politics* and *American Journal of Political Science*]. . . . They had decided this was political science, and would not let biases of others interfere and papers would be validated on their merit."[15] Many others cited other generalist journal editors who informally voiced support for the new field.

But researchers also credited editors of specialty journals who have supported their research, editors like Michael Lewis-Beck of *American Voter* or Bob Jervis at *Columbia International Relations*.[16] From the perspective of social genomics researchers, the key to positive reception in journals has been having key senior "untouchable" scientists on editorial boards who can stick their necks out and argue for a fair editorial process.

Even with open-minded editors out there, review panels are still a significant issue in the publication of social genomics findings. Social science editors face the challenge of assembling diverse enough review panels to judge social genomics work. With papers that may have twenty to upwards of three hundred authors on them, all with different skill sets who are reporting in detailed ways on social behavioral elements as well as genomics, it is difficult to find external reviewers who single-handedly embody the span of expertise needed. It's just as hard to find the right combination of specialists or to get reviewers on who aren't already somehow linked to the papers. And as one researcher put it, "One or two people could do [one study] but they might produce one article every eighteen months."[17] Teams are an essential apparatus of social genomics research, and thus are nonnegotiable.

That's part of the reason why most researchers say that only really big science journals are equipped to publish their work in an efficient and timely manner. Time to publication is the single biggest challenge with social science journals. As one researcher complained, his two earliest research papers in the area took years to get accepted.[18] One was rejected by the top economics journals, including one where an editor informally told him that the journal just wasn't ready for a gene-environment paper. The second paper did eventually reach a top economics journal, but only after six years and six rejections. In other words, the field had to catch up substantially before the paper could find a home.

But even now that leading journals are enthusiastic about social genomics, garnering first reviews in a social science journal can take a year or more. As one political scientist who has published in the field's top journals put it: "We published fairly consistently in political science journals. We've just found it easier to publish outside of political science than inside of it."[19] Or as a sociologist who has published in his field's top journals has said: "If we want to keep up with the new techniques on everything, we should publish our work fast, because this method may be attractive this year but not in five years. Some people may find this out of date."[20] One economist who was invited to publish in a leading journal and was assured that his paper would be "fast-tracked" ended up mired in a four-year process with one stubborn reviewer. As he lamented, there were three or four rounds of reviews and "every round was very time-consuming because we had to do a lot of science research and talk to people in genetics, asking, 'Is this claim true?'"[21] The bottom line is, "if you have a hot genetic finding, you are simply not going to wait the year to several years it takes to publish it in the *American Economic Review* or the *American Sociological Review*."[22]

Researchers see publishing in natural science journals as a no-brainer. Those working with genetic experts in other fields, such as clinical psychologists and medical specialists, say that their collaborators' home fields just look at the genetics in their papers as cut-and-dried methodological issues. Besides, behavior genetics journals and other journals partial to genetics have reached out and made themselves available to these interdisciplinary teams.

All this boils down to the fact that different standards abound across journals. *Science*, for example, publishes studies that follow traditional neuroscience design standards of recruiting less than one hundred people as well as large-scale studies on hundreds of thousands of people. Journals that do not publish such a range of study designs often do not recognize the range of social genomics studies as science. In addition, some journals are more interested in methods, and so are more open to un- or undertheorized work. For example, within the social sciences, journals like *Demography* have been more willing to publish ethically controversial work as long as there has been good methodological discussion. Meanwhile, social science specialty journals continually ask researchers to build out theory in their domain. One research team reporting the results of a candidate gene study reported facing this challenge until it went to the PLOS journals, which had no standard metric for evaluating the research beyond genetic methodological validity.[23]

Fervor from Funders

Social genomics, and sociogenomic science more broadly, is what it is because of its high-profile publications but also its warm reception by the international funding communities. Funders public, private, and everything in between have been excited to help this new scientific framework take off.

Several major public agencies are go-to funders for social genomics researchers working worldwide. The first is the NSF, with its Directorate for Social, Behavioral and Economic Sciences. Within that directorate lies the NSF Division of Social and Economic Sciences, which is focused on funding research into "human, social and organizational behavior by building social science infrastructure, by developing social disciplinary and interdisciplinary research projects that advance knowledge in the social and economic sciences." It is explicitly interested in social genomics because of its premium on interdisciplinary research.[24] The agency makes clear that it will facilitate joint review among Social and Economic Sciences programs and programs in other divisions, as well as NSF-wide programs for projects that encourage cross-pollination among the sciences. The NSF is also particularly interested in "the intellectual and social contexts that govern the development and use of science and technology"; therefore it prioritizes social genomics proposals for moving technology forward throughout the sciences.

Within the NSF Directorate for Social, Behavioral and Economic Sciences, social genomics researchers can also apply to the Behavioral and Cognitive Sciences division. While the Social and Economic Sciences division has been the usual place for economists, sociologists, and political scientists to get NSF funding, with the new emphasis on biology and genomic methods researchers are now highly attractive to the Division of Behavioral and Cognitive Sciences, which emphasizes its interest on "biocomplexity and the environment, children's research, cognitive neuroscience, human origins, and information technology."[25] Even though the Division of Behavioral and Cognitive Sciences traditionally funds research focused on "human cognition, language, social behavior and culture, as well as research on the interactions between human societies and the physical environment," studies on the range of behavioral phenotypes, such as aggression, educational attainment, and violent behavior, all relate to cognition and social behavior and are thus fair game.

The agency has also sponsored an annual international workshop on statistical genetic methods for human complex traits targeted at political and social scientists. This five-day advanced course, which takes place at the University of Colorado's Boulder campus, is a fundamental training hub for sociogenomic science.

Because the NSF supports such a wide range of scientific activities—everything from original research to training programs, workshops and conferences, and the build-out of research facilities—it is an important source of funding for the emerging field. NSF-funded social genomics projects include John Alford, John Hibbing, and Kevin Smith's 2007–10 "Genes and Politics: Providing the Necessary Data" and "Investigating the Genetic Basis of Economic Behavior," and their 2009–12 "DHB: Identifying Biological Influences on Political Temperaments";[26] Quamrul Ashraf's 2013–15 "Genetic Diversity and the Wealth of Nations";[27] Christopher Dawes's 2010–11 doctoral dissertation research in political science, "A Genome-wide Association Study of Voter Turnout";[28] James Fowler's 2007–10 "The Genetic Basis of Social Networks and Civic Engagement";[29] and Jamie Settle's 2014–16 "Understanding the Mechanisms for Disengagement Contentious Political Interaction."[30]

The NIH is another primary source of funding for social genomics research being conducted in the United States as well as abroad. Within the NIH, the National Institute of Child Health and Human Development offers meeting grants and project grants for "biomedical, biobehavioral, and interventional research that improves the health of people throughout the world, including vulnerable populations such as those with disabilities and those in resource-poor areas."[31] The institute is particularly proud of its having funded genetic discoveries having to do with developmental disorders such as Fragile X syndrome and Rhett syndrome, diseases for which there are now prenatal genetic tests. Like the NSF, it is also committed to fostering interdisciplinary research.

It offers a Program Project Grant that funds multidisciplinary science teams that combine three or more child and development projects.[32] Examples of funding from this institute include Jason Boardman's 2009–10 "The Social Determinants of Genetic Expression: A Life-Course Perspective,"[33] the 2010–15 "Social Demographic Moderation of Genome Wide Associations for Body Mass Index,"[34] and the 2013–15 "The Social and Genetic Epidemiology of Health Behaviors: An Integrated Approach";[35] and Jason Fletcher's 2012–15 "Examining the Sources and Implications of Genetic Homophily in Social Networks."[36]

The other main institute at the NIH that teams can seek funding from is the National Institute on Aging, which supports a critical component of social genomics research: studies on adult development and changes through the life course. Cognitive function and performance, health and chronic disease, and behavior as it involves adults moving through life are all foci of this funding department. Funded projects include Nicholas Christakis's 2008–9 "Networks and Neighborhoods," "Genes as Instrumental Variables for Health Behavior Peer Effects in the FHS-Net," and "The Genetic Basis of Civic Engagement and Social Networks and Their Effect on Health";[37] Jeremy Freese's 2015–18 "A Longitudinal Resource for Genetic Research in Behavioral and Health Sciences";[38] and Michael Shanahan's 2010–15 "Genetic Risk, Pathways to Adulthood, and Health Inequalities."[39]

At the NIH, researchers are also encouraged to take advantage of the Division of Behavioral and Social Research funding, which more generally supports social, behavioral, and economic research, including its own R01 programs, the 07-070 and 11-260 research project grants. Funds from the Division of Behavioral and Social Research are intended to bridge disciplines and institutes within the NIH—in their words, "cross-disciplinary research, at multiple levels from genetics to cross-national comparative research, and at stages from basic through translational."[40]

In Europe, social genomics researchers of varying nationalities have looked to the European Union for funding. In particular, they have found success with the European Research Council's Consolidator Grant mechanism. Consolidator Grants are designed to fund junior scientists (scientists seven to twelve years out from their doctoral programs) to become successful autonomous investigators, and to fortify emerging research teams that hang in the balance. While Consolidator Grants are not specifically targeted at gene-environment scientists or interdisciplinary researchers, they are designed to support the building of labs that go beyond the demands that a typical university addresses.

Consolidator Grants, more than anything, are aimed at supporting the hottest new research areas such that the world's most innovative scientists (hailing from non-European as well as European countries) will stay in Europe and provide training to European students.[41] Social genomics has been attractive to the European Research Council for this reason. Consolidator Grants have gone to researchers like Melinda Mills and Philipp Koellinger, for Mills's 2014 "Unravelling the Genetic Influences of Reproductive Behaviour and Gene-Environment Interaction"[42] and Koellinger's "The Molecular Genetic Architecture of Educational Attainment and Its Significance for Cognitive Health."[43]

A less obvious public source of social genomics funding concerns the military. Just in the United States, social genomics finds support from the Department of Defense, the Army, and the National Institute of Justice. The Department of Defense provides original research support through its Office of Net Assessment Basic and Applied Scientific Research Grant. This mechanism funds projects that "compare the standing, trends, and future prospects of U.S. and foreign military capability and military potential."[44] National security research is the agency's main aim. Meanwhile, the Army offers R01 funding in conjunction with the National Institute of Mental Health, and the National Institute of Justice offers graduate research funding with a fellowship program. These sources of funding sponsor research into mental health, suicide, and crime and justice in the United States. Military funding has supported research like Kevin Beaver's 2005 "The Intersection of Genes, the Environment, and Crime and Delinquency: A Longitudinal Study of Offending";[45] Rose McDermott's experiments on biology in international relations;[46] and Colter Mitchell's "Army STARRS Statistical Design and Methods for Integrating Biomarker and Genetic Data into the Army STARRS Epidemiological Study of Suicidality."[47]

A number of foundations also fund social genomics research. Some, like the Russell Sage Foundation, hold multiple mechanisms for funding original research. Russell Sage has three relevant core funding opportunities: the Program on Social Inequality, the Behavioral Economics program, and the Future of Work program. The Program on Social Inequality funds projects on economic inequality broadly speaking, including research on American social institutions, social mobility, social opportunity, and intergenerational transmission of advantage.[48] Its Behavioral Economics program funds behavioral research on matters like "poverty, inequality, intergenerational mobility, labor markets, public finance, parenting and child development, and racial and ethnic bias" that can shed light on and possibly improve American social and living conditions.[49] The Future of Work program funds research on employment, including job quality, labor policy, employer practices, and earnings. While none of these mechanisms seem directly connected to social genomics, each has an area that it is particularly excited about funding. The Program on Social Inequality wants to learn new things about intergenerational transmission of wealth; Behavioral Economics seeks new models of economic decision making and behavior beyond the narrow rational choice theories; and the Future of Work is "especially interested in proposals that address important questions about the interplay of market and non-market forces in shaping the wellbeing of workers, today and in the future." Intergenerational transmission, decision making and behavior, and nonmarket forces shaping the future all resonate with the genetic explanations social genomics gives for social interaction, status, and processes.

In 2016, Russell Sage issued a fourth mechanism: its special initiative on Integrating Biology and Social Science Knowledge (BioSS). As it states: "This initiative will support innovative research that incorporates biological concepts, processes and measures in social science models in an effort to improve our understanding about questions of interest to RSF's core programs." The foundation lists topics like immigration and social mobility as critical to its mandate. It is important to note that its board includes prominent social genomics researchers. Russell Sage has funded projects like Dalton Conley's 2015 "GxE and Health Inequality over the Life Course";[50] Chris Dawes's 2006 "Role of Genetics in Preference Formation";[51] a conference and special issue of the foundation's journal on biosocial pathways;[52] as well as a Summer Research Institute in Social Science Genomics.[53]

Other foundations support original research projects through the funding of an individual's career. The Guggenheim Foundation, for example, offers the Guggenheim Fellowship, a midcareer award for people who have "demonstrated exceptional capacity for productive scholarship."[54] Guggenheim funds hundreds of researchers each year in the Americas. Because it doesn't have a specific set of research funding aims but rather is career based, it is wide open for those building out the field. Guggenheim fellows have included Dalton Conley, for his research into "genome-wide studies of social outcomes to better determine the heritability of behaviors."[55] The William T. Grant Foundation, which supports research into youth, also supports social genomics research through its Faculty Scholars career development program. As with most career development grants, it supports a promising researcher to establish an innovative research program and to develop new dimensions. Its focus on youth includes child development, education, and inequality. The Grant Foundation has sponsored Jason Fletcher's 2012–17 "Interconnected Contexts: The Interplay between Genetics and Social Settings in Youth Development" and Joshua Brown's 2014–17 "The Impact of School and Classroom Environments on Youth Mental Health: Moderation by Genetic Polymorphisms."[56]

There are also funding opportunities at specific universities, such as the University of Michigan's MIDUS Grants and Population Studies Center Research Project Funding, or Columbia University's Institute for Social and Economic Research and Policy and Population Research Center Grants. These awards fund new avenues in demography. Such awards have gone to scholars like Jason Boardman for his 2006–7 "Genetic Aspects of Psychological Resiliency among US Adults" pilot study;[57] Colter Mitchell for his 2014–16 "Applying Whole Genome Data to Common Social Science Issues";[58] and Jason Fletcher for "Exploring the Relevance of Gene-Environment Interactions for the Social Sciences."[59]

Finally, there are nonprofits that support research working groups, conferences, and symposia, such as the National Bureau for Economic Research, the largest economics organization in the United States. It sponsors collaboration between economists and social scientists who are researching new topics in the areas of aging, asset pricing, children, corporate finance, development of the American economy, economics of education, economic fluctuations and growth, energy and the environment, healthcare, health economics, industrial organization, international finance and macroeconomics, international trade and investment, labor studies, law and economics, monetary economics, political economy, productivity, and public economics. The bureau has funded the work of researchers like Daniel Benjamin, Dalton Conley, and Emily Rauscher, including Conley and Rauscher's 2010 development of "Genetic Interactions with Prenatal Social Environment: Effects on Academic and Behavioral Outcomes." National Bureau for Economic Research funds have also contributed to SSGAC study meetings outside the CHARGE conference framework.[60] Professional associations also fund meetings and workshops to encourage novel research agendas that they believe will come to matter to their constituents. The Population Association of America has been one of the main funders of the IGSS, while the American Sociological Association and the Royal Economic Society have hosted a number of methodological workshops to spread the use of social genomics approaches.

"We are after all biological": Social Science Funders Go Sociogenomic

What's most obvious from looking at the list of *who* funds social genomics and *why* is that social science funders are the real muscle. But there is more to the story than a simple catalog can tell. I had to meet and speak with funders to really understand why millions are being invested in the new science. What I found was that these social science funders are extremely enthusiastic to support social genomics because they are excited about biology. And they are eager to create more mechanisms for sociogenomic support because they believe that the science will have important social and policy implications in the years to come.

Funders are so fervid that they are hands-on with the science. They come to meetings, debate techniques, and interact with cohort studies to facilitate data support. They express the sense that the need between social genomics researchers and social science funders is mutual. As one NIH funder put it, "If you're serious about society, . . . we are after all biological."[61] I heard this rationale from just about every funder I spoke with.

Indeed, funders' excitement about the new field is evident in their descriptions of it. Recall the statement from the NSF official in Chapter 2. She said that social genomics is "high risk, high reward" and "could be potentially transformative, . . . exactly the kind of work the NSF seeks to fund."[62] A sponsor at the NIH said that the field was a long-needed relief from monodisciplinary work, "these really slow-moving boats between sociology, demography, and economics and behavioral genetics." Social genomics was finally "coming up with common language and common agreements."[63] Another sponsor at a different NIH institute talked about social genomics research as a first step to a broader interdisciplinary research agenda. When I asked if their focus on GWAS was cutting-edge enough for the agency, she said:

> Actually, the associations and the correlations will never lose their value in terms of showing people where to hunt for mechanism. They have that role to play and that's the thing. The mechanistic work is so particular that if you don't have a broader review of where to go hunting, you'll get lost. So yeah, that work needs to keep being done but it needs to be framed properly. It needs to be framed as showing you where to go looking for deeper studies.[64]

This funder talked about the difficulty of working on innovation around gene-environment research with national and supranational bodies like the UK's Medical Research Council, year after year. As she put it, "Everybody needs to be working to figure out how to handle the speed at which [the train's] moving, as much as how to handle finding the correct models."[65] Social genomics seemed to bring many loose threads together in ways that capitalized on their different strengths.

The field's interdisciplinarity is the main attraction for funders. One foundation rep lauded the field for taking the plunge into biosocial data: "Given the fact that so many data sets are now collecting genetic information and other forms of biological information, there is a role for better understanding how biology can inform social science research."[66] The NSF director above echoed this rep's plaint that interdisciplinary research had been on the mind of funders for some time. She said this interdisciplinary orientation had been the agency's priority for quite some time:

> The broader issues of how family background affects social and economic outcomes have just been really fundamental to the social and behavioral sciences for a really long time. And I'm saying family background because for many people that includes, depending on whether or not you were raised with your biological parents, . . . everything from genetic, prenatal exposure, family environment, after birth, yadda yadda yadda. In that sense, it's not a brand-new set of things for us to be thinking about in any way in our sciences. It is a further development given developments in genetics over the past—God! I know somebody from grad school who was doing early work in genomics in the 1980s, okay? So I don't want to make it sound like this is just something that blew up out of nowhere, because it's been around for—it's taking awhile to get into the social science community, but it's making those linkages.[67]

Funders shared their personal dreams of how the sciences could transform. As one NIH funder exclaimed, "Can you imagine a day when sociology departments actually also have a biological curriculum and like labs and stuff?" He then listed several leading social genomics researchers and said, "Among the intellectual leaders and the people who are trying to move, it's like, it's not *either*. It's *both*."[68]

This funder said the crisis in scientific understanding, the very crisis that requires bridge builders like social genomics folks, comes from the rift in the natural and social sciences. As he described it, there are two camps: social scientists who are interested in power and are tired of everything being sold as genetic, and geneticists who don't want to be reductionist and therefore ignore social traits. To him, social genomics was the future, or "the huge opportunity of tomorrow where stats and biology come one in one."[69]

In fact, the funders I spoke with believe that the new field is better poised than other tried-and-true fields to do this interdisciplinary work. As the NSF administrator told me, what she fears most about gene-environment science is "junk science," a problem she sees as common in interdisciplinary endeavors where mainstream biologists are leading the way:

> BioMed doesn't work traditionally with—there are real issues. In fact, the reason for the consortium's existence was the realization that it was going to take a really big dataset to find anything that was reliable. So you're at the risk of people looking at just, "This is the data," and going, "Oh my god! We found the gene for such and such," where all they had is just a statistical artifact.[70]

Reflecting on social genomics and its innovative approach, she continued:

> They bring together the large sample sizes, the pulling of the cohorts studies, and they bring together those people with backgrounds in genomics, but also people who have thought through carefully how you're going to measure some of these social or behavior or outcomes variables. You can use education and intelligence. You can use intelligence as an example. It sounds quite easy to say, "How do we measure intelligence?" It turns out that's actually way more complicated than people realize. . . . More generally, I wouldn't expect a group of medical researchers to have thought as carefully about how to measure social and behavioral outcomes or the complicated processes that lead to those outcomes as people who have trained specifically to think about that.[71]

Her take was that social scientists know best how to study humans, not just animal models and cellular mechanisms but the whole human. Or as one NIH official put it, "You still have a lot of entrenched methodologies . . . in part because people who are now doing genetics didn't really come up from looking under a microscope."[72] Social genomics, with its capacity to study gene-environment interactions in a holistic way would do richer inquiries than traditional genetics and behavior science.

As a result of this confidence in social genomics, some funders have been attempting to create more opportunities for support. One program officer told me that he was working on putting a funding mechanism together just for this kind of research.

Our board authorized funding to support the activities of a working group to sort of help us think about where we might be able to make an impact in this area. Clearly there's a lot of money out there available and a variety from the Feds and other foundations. The question is, for a social science funder like us, where might we be able to help establish a research agenda and move the field on forward especially for social scientists. At this stage, we are still in the early part of it but I would expect to see us doing more probably within the next year.[73]

Indeed, within months his foundation had exponentially grown its sponsorship and role in the field, offering conferences and training forums.

In addition to speaking with funders, I also spoke to cohort study leaders and biobank administrators about their involvement and support of social genomics. They exclaimed the same eagerness and enthusiasm, and even a lot of the same views. As one leader of a large-scale US cohort argued, monodisciplinary genetics was probably worse off than multidisciplinary social genomics:

Just in the sense of social scientists, [where] our technology will make mistakes with genetic data, so do geneticists who aren't that familiar with social science data make mistakes. Like they end up with, because they don't have good data and so on, end up with a sample that's one-eighth the size of what they really could be using.[74]

Social genomics with its greater attention to statistics would bring about a "greater interdisciplinary presentation of these results," which would lead to a better "training of new users" and better research on the whole.[75]

Others spoke about this in terms of assessing the gene-environment relation. A leader of a population-based biobank said that social genomics was better poised to study gene-environment interactions because of how broad the environment really is:

The environment being, oh my goodness, many, many things, including the physical environment. So that could be the air pollution of the neighborhoods that are poor. I mean this sort of broad exposures. It could be the built environment, something about the way the neighborhood was constructed, public housing, and things like that! It could also be policy. I mean, policy is an environment.[76]

This study administrator felt that social genomics researchers would be able to tell "whether certain interventions are going to be more effective, *social* interventions are going to be more effective, depending on whatever your sequence is." Social genomics researchers were getting into "the middle ground" between genetics and social science "where both sides can play together."[77] An American cohort study leader similarly said:

> Our major perspective is that you have to know a little bit about—you have to collaborate or know quite a bit about molecular genetics, cells, but also, really think about how to measure environmental types or enviro-types, which measure the environment well. I think one of the major challenges that this area faces is the measurement of environments and phenotypes. You'll notice at those meetings, you have some molecular geneticists, not too many but some. Then you'll have a whole lot of social scientists who work at a fairly—the level of data that they work with is fairly high.[78]

But cohort study leaders and biobank officials are also supportive of social genomics because they believe that the social science bent of the field is more likely to produce ethically responsible science. As one European cohort study leader said:

> We need to be a very heterogeneous group because we need knowledge from different disciplines. . . . We definitely need more [of] those scientists and other social science people, because there is—the more we get these results of social behavior and genetics out, the more the question of how do these results relate [to] political atmosphere, political decision making, is bound to come up. We definitely need to be prepared for answering because there is a risk that these results can be taken in ways that we never wanted them to be taken. Say, for instance, if their education study had shown up that yes, there is a clear gene related to educational attainment, then there is a big risk that would be taken in such a way that yes, people can be divided by genes into different groups. Who gets to go in the best schools and who doesn't?[79]

This cohort study leader said that this dilemma "is just behind the corner," something that only "genuine social science people" could manage.[80]

What Other Scientists Think

Hearing the celebratory views of institutional sponsors made me wonder if the entire science community stood behind social genomics. In fact, it's a mixed bag. There are many social genomics collaborators, and many firms and organizations that are partnering with social genomics researchers to make studies happen, but there are also some detractors of the science spread throughout the sciences. Let's take a look at their perspectives.

At conferences and study meetings, I met a number of collaborators and supporters who echoed the views of funders. Some provided DNA samples or analysis to researchers. Others developed theory to move the field along. Still others were the glue or go-betweens that link researchers to databases, cohorts, or research support. These scientists naturally believed that the science was robust and the findings valid, and many were committed to continue edifying the field's standing in their own disciplines.

One well-known behavior geneticist who has served as past president of the Behavior Genetics Association told me that he was involved with social genomics out of a mutual need. They need him because "quality control is a real issue . . . you need to have a very solid training in this stuff and know things that you don't want and what the traps are." But he also needs them, because social genomics helps to validate the science he has done for decades:

> Occupational stages, entrepreneurship, all these things, I mean we and others have published twin studies showing that there are genetic components for these things. And I just think it's going to really—in a sense, all the stuff that it's doing is confirming what we've learned from twin studies for ages.[81]

This scientist has been especially supportive of social genomics because he believes that as an unfettered new science with no bad track record to speak of, it will do a better job of persuading the sciences and the public that there are genetic causes for human behavior.

Another leading behavior geneticist of international renown agreed. In our conversations, she emphasized the ways in which social genomics extends the work of prior scientists:

> So from our perspective we have always worked on these traits, so we don't have any issues applying newer genetics methods to these traits, which has been a research of over forty years. The concerns are in making sure that the analyses and the research is done correctly, and making sure that the individuals who are new to the field regardless of what traits they work on have access to experienced methodologists to make sure that the work that has been done is being done correctly.[82]

This scientist assured me that the only difference in behavior genetics and social genomics is "in the interpretation of findings" which is guided by the different theoretical training and interests different fields hold.

Even scientists who have had no involvement with the field back up the science and voice support for it. One scientist at BGI's Cognitive Genomics branch, home of the much maligned IQ genomics project, told me that social genomics is pretty much the same as what they do at BGI, save for study design:

It's almost identical to what we have been doing in our lab at BGI, just that their population is a little bit different and their phenotype is slightly different. Because the main thing they can access is years of educational attainment, although on some subsets of their data they actually have IQ scores and we have been a little bit more focused on IQ scores of those actually qualified for our study, some of the people qualified on the basis of having a PhD or something like that from an elite institution, so even in some sense we were using educational attainment too.[83]

This scientist applauded the "random population" design of the field's genome-wide association studies, saying that it's a direction that BGI is interested in taking too. He also applauded the field's strategic partnership with the right people in genetics and genomics:

If you look at SSGAC, you can look at their supplements, and they are following the best practice guidelines that come out of places like the Broad and others. Also Peter Visscher, who is one of the coauthors on SSGAC and actually one of the leaders of SSGAC, is actually, you could say, the single-most—I would say personally, he has made the single biggest contribution to the field of genomics, broadly speaking, in the last decade or so.[84]

This scientist expressed the belief that social genomics is just carrying out the mandate of basic science:

It's basic science to just know whether the characteristic of a person—to what extent it's determined by DNA. And which particular chunks of DNA are contributing to that variation. It's just a basic scientific question.[85]

He added that GWAS was the "gold standard itself," therefore, "unless you are really a Marxist or Leftist" and you don't believe that "the construct called cognitive ability exists," you would have to agree that social genomics is taking science in a good direction.[86]

The behavioral scientists I talked with spoke highly of social genomics, but I found an almost polar opposite response from mainstream genome scientists. One leading genome scientist who has published some of the field's most influential neurogenomics papers, a scientist who has recently served as president of the American Society for Human Genetics, said that the field's approach was nonsense. He was especially concerned about the ways in which social genomic researchers were characterizing social phenomena as medically relevant traits, as well as their use of cohort study samples without proper consent.[87] This scientist quipped, "If one is interested in studying the genetics of socially determined traits, it might be more lucrative to study the genetic characteristics of those who are creating the social environment that determines those traits than those who carry them." But despite his criticism of the science, he didn't have any plans to shut it down. He hardly believed that any serious genetic or genomic scientist would take social genomics seriously.

To his point, there has only been one public response from the genomics community. In 2010, a group of molecular scientists, evolutionary biologists, and anthropologists at Harvard spoke out in opposition to "The Out of Africa Hypothesis, Human Genetic Diversity and Comparative Economic Development," the article that argued that societies with genetic diversity were economically fitter.[88] This team of scientists wrote that the paper was "deeply scientifically flawed" with "highly tendentious implications." They argued that

> Ashraf and Galor misuse genetic, evolutionary, archaeological, historical and cultural data to propose that there is some optimum level of genetic diversity that favors economic development. They argue that very high genetic diversity in a given society engenders greater distrust and reduced cooperation, while societies with low genetic diversity lag in innovation. The argument is fundamentally flawed by assuming that there is a causal relationship between genetic diversity and complex behaviors such as innovation and distrust. The argument is also statistically flawed by treating genetic data as each population having an entirely independent history both from a genetic and from a historical point of view, when in fact, they are highly correlated and inextricably entangled with genetic population structure and with contingent historical events. Such haphazard methods and erroneous assumptions of statistical independence could equally find a genetic cause for the use of chopsticks.[89]

Though this critique was taken up in the scientific media and translated for wider public notoriety, as a *Nature* reporter noted, "The manuscript had been circulating on the Internet for more than two years, garnering little attention outside economics"[90] until *Science* posted a blurb on the study.[91] After its authors reinforced their position in interviews, citing a "misuse of data" that could lead to racist interpretations and social implications, the group went silent again and has remained so since.

I asked genome scientists to tell me what it is about the field of social genomics that concerns them. One scientist focused on epigenetics said:

> It's very distinct in lots of ways from the kinds of science that I do. In fact, my own research, which is largely in epigenetics, it also focuses on sort of the entanglements between social and biological factors, but in a different way. It's explicitly intended not to sort of be overly reductive in the way that I see a lot of the social genomics work to be situating itself. Social genomics is sort of a misnomer. I don't see anything particularly social about social genomics, because there is very little emphasis on social factors and there tends to be this really large emphasis on biological ones.[92]

This epigeneticist sees two major conceptual problems with the science of social genomics: "slippage between what is biological and what is genetic, and how each of those levels sort of articulates and entangles with social factors." As he remarked: "Biology obviously involves a wider spectrum of much higher level and more complicated processes than genetics does, and often we conflate those two terms. I see that problem a lot when I read stuff that comes out of social genomics, and that gets magnified in media coverage."[93]

Like many others I spoke with, he worries that these flimsy characterizations will intersect the field's haphazard explanation of the environment. As he put it, "people have agency" and behavior is "extremely plastic and flexible."

> So because the trait events are often complex, the genome-wide association correlates may be the result of some like underlying factor or a separated factor that's correlated a sort of messy proxy that we have for the trait that we are interested in. That means there is a real lack of precision from the get-go in defining the behaviors under study which are already complex behaviors, which on top of that are cultural and constituted behavior.[94]

In addition to these issues, this scientist sees the statistics and evolutionary logic as bunk. In terms of the latter, the field just generates a bunch of genetic markers that "don't actually ever directly explain sort of the genetic architecture of the trait in a specifically causal biological events." As he put it:

> That SNP itself is not a gene. It is a marker that may be close to a gene that might have some significance. . . . Without an explicit functional link, like this gene's involved in cognition in this region of the brain because of this, and it varies in

this way by these mechanisms— . . . if you can't give a specific functional link there is no reason in even thinking that there is a biomedical truth or intervention that can be made if we can't understand the functional relation.[95]

Statistically speaking, he takes issue with the very kind of methodological innovation that the field has made its name by: large-scale multi-cohort GWAS.

> I mean it seems impressive, but the reason that it is (and it is impressive in the sense that that's a lot of data), the reason that it's so large is that the effect size of the loci that we are looking for, these genetic influencing factors—the reason that they are so large is that the impact of those genes is so tiny that the sample size has to be huge to find a sense of that. . . . GWAS has a really big problem with missing heritability. So I think social and environmental factors should be carefully considered and not sort of given lip service.[96]

He called educational attainment "a proxy of a proxy of a proxy of a proxy, that cognitive function of an already complicated polygenic trait in itself." Without an explanation of how genetic variance explicitly adds months of educational attainment, he wasn't buying it.

Another expert who contributed to the 2010 critique said:

> Any sort of behavioral trait is only partly influenced by genes and even in that case there are likely to be many thousands of genes involved, and the genes involved are likely to have to be different depending on environmental context, but they have a different structure of relationship to the trait. We also know that in the behavioral sciences the traits under investigation are very theory laden and socially inflected, so something like voter behavior is a funky kind of phenotype to be investigating. So there has to be significant sort of presumptions of an endophenotype underlying or maybe multiple endophenotypes underlying some sort of construct that they are looking at. That is to say these studies don't avoid either the classic problems of defining phenotype in behavioral studies or the classic problems of dealing with large numbers of gene variants.[97]

From where this researcher is seated, the trademark social genomics GWAS method looks "really old-fashioned and they look disengaged with the cutting edge in gene environment," such as epigenetics, dynamic genomes, and systems biology.[98]

A different critique of social genomics comes from scientists who have tried to do gene-environment studies not so different from social genomics, and who have turned up cold. Remarks from one genetic epidemiologist exemplifies this perspective. This scientist had stimulus support from the NIH that has been followed with R01 support "to develop a good gene measure that would be predictive of [the social behavior under study] and then integrate that into G-by-E models of the development of [said social behavior]." Yet as he said:

> We are entering year five now of the R01 and we still don't have a reliable gene measure that we can pull into our models. It's not just us. It's a common problem that's going on in my field. We do have a number of candidate genes that we have tested that are often cited as good candidates for [the phenotype in question]. But the paper we're writing on that is a null findings paper. Across four longitudinal studies, we've got nothing. And this is for like dopamine, DRD2, DRD4, MAOA, COMT, and serotonin transporters. We have no associations. . . . We also tried developing gene system SNP scores within GABA and dopamine and serotonin systems to come up with a cumulative SNP index. None of those were predictive.[99]

This researcher, like others who have come up null, describes the field as full of false positives and unreplicable findings to the point of being unscientific. He said people just keep telling him "the signals are there and it's just more complicated than we thought it was and it will be a little harder to find, and once we got consortia with big enough data sets and this index, this will be developed." But he is skeptical. He feels that the missing heritability might be found in the epigenome. Or it might just not be there at all.

He still holds out hope for adding individual variation to biosocial models:

> This would push prevention science to pay attention to individual differences in kids and they may not be genetic, but they could be of social origin as well, these differences. But that's why I got on board is because I felt, "Well, let's look at the individual differences for a few years and see how do they differentially interact with environmental experiences." So that's on the positive side why I did decide to go along with it. But I still have reservations especially in view of the fact that, well, like the Duncan article and other articles that have come out of my field, suggesting that a large proportion of the last ten years of G-by-E research has been false positives. I wonder what happens if the popular press catches onto the original paper that comes out and says, "MAOA and maltreatments as an interaction," but they never come up with the "Oh, but people failed to replicate that."[100]

In other words, the prospect of media buzz around inadequate work only se-
cures his skepticism of any sociogenomic research enterprise.

The last round of criticisms I encountered comes from social scientists who
have worked in biomedicine and who are looking on at the field of social genomics and reflecting on what good science means to them. One sociologist
who calls for responsible uses of "biology in service of sociology rather than
sociology in service of biology" said:

> I noticed that a lot of the papers were heavily methodological, and this seems
> to be without a lot of attention to issues that are or should be of fundamental
> importance to sociologists; things like gender, like race, and when these things
> are discussed or operationalized it's often in a pretty simplistic way.[101]

This researcher voiced the sentiment that the field is more interested in advancing methodology, "or to advance the biology really, or the demography,
or maybe to help research," and less in advancing the fundamentals of social
science.[102]

Another criticized the "teeny tiny effect sizes" of social genomics findings,
asking why anyone would want to pursue this line of thinking:

> The level of significance that many of them are able to attain is incredibly small.
> It's possible. I would grant that probably there's something in our DNA that
> somehow runs through some sort of pathway which we can't identify yet all the
> way out to social behavior; like I would be fine to grant that. But I think that
> the amount of influence the genome has is vanishingly small for most of these
> behaviors and to spend a lot of money and a lot of time looking for those incred-
> ibly small effect sizes is a waste of money and it's a waste of time and draws away
> from what I think of as much more effective research questions, and as a result,
> policy solutions.[103]

This social scientist has personally expressed dismay to funders interested in
creating funding mechanisms for social genomics, telling them that "there are
things that we know affect educational attainment like not having any food
in your belly . . . or going to a really crappy school where the teachers don't
get paid enough." She prayed that scientists would stop "chasing in this shiny
genetic explanation that remains shiny even though we now know how many
other things matter," and instead focus on actionable biosocial interactions. As
she relayed:

In the world of medical genetics which is actually the research I was doing . . . it's much more clearly delineated. You have a disease or you don't. You have a SNP and it raises your risk or it doesn't. And yes, there are so many cases where they find SNPs that later they find are actually associated with that disease. But even in that world of medical genetics, they don't have a clear idea of the pathways through which that SNP actually results in disease or how much risk it confers, or the relationship of SNPs to each other. So there are still so many unknowns when it comes to medical outcomes. And that's why I just think when you shift your perspective then to social outcomes which are that much less clearly delineated and are that much harder to think about—like what the pathway would be—then I would be a little bit concerned about trying to think about policy approaches to dealing with them. So let's take a hypothetical here. So, if you said to me, "*Nature* is getting ready to publish the [educational attainment] paper where there are seventy-five SNPs, and all together those SNPs suggest that they're explaining fifteen percent of the variation in the educational outcomes," that's a hypothetical that's way more of a [cheap policy] spend than what's really happening. What do we do with that information? Would we screen intensively at birth and put [kids] into different schools where they were likely to have teachers who knew how to deal with people without these SNPs? That seems like one possible policy approach which would be like enormous amounts of overkill given that so many other things like poverty and parental education and quality of school affect educational outcomes so much more. When we start spending all of our time doing genomic screening of babies, so we try to deal with that fifteen percent rather than building better schools and paying teachers more and trying to get food and good values of poor kids, this is where it's like even if you've told me that the state of the science was that actually they can explain some significant minority of the variation, I would still say there's a lot of other policy approaches that we know would work and that we know have a big outcome.[104]

For social researchers like this one, focusing on "society and humans and nongenetic factors" that are known to produce the vast majority of variance for behavior and outcomes seems a better approach that still has not been pursued to a satisfactory degree. In this example of educational attainment, this expert expresses a common fear that policymakers will be seduced by easy outs from the social structural changes that often are harder and more costly to make.

Despite these well-thought-out criticisms, I must emphasize that I encountered far more social scientists who were enthusiastic about social genomics.

Noting how few natural scientists had anything bad to say about social genomics in the press and privately, I could only conclude that the science community is approaching sociogenomic science with arms wide open.

Disciplines for Interdisciplines

With such minimal opposition, and a wealth of support in publishing, funding, and the greater science community, social genomics and its sociogenomic framework are thriving. But it would be remiss of me not to mention one other area of support, especially since this area is one of the most critical bases for the existence of new sociogenomic sciences. That is home disciplines and home departments. As you've already seen, when talking with social genomics researchers about where they have felt support or pushback, departments and disciplines were at the top of their lists for both. My analysis of actual productions put on by disciplinary entities only attests to the avid interest they have in promoting sociogenomic science.

One way to witness the support of disciplines and departments is through the activity of professional associations. In 2015 alone, the American Sociological Association, Royal Society of Economics, British Sociology Association, American Political Science Association, Population Association of America, and American Society of Criminology all held panels on social genomics. With titles like "Individual Differences and Criminal Behavior: A Focus on Environment, Molecular Genetic Variation, and Psychopathy," "Genetic Risk and Family, Environment and the Life Course," and "The Role of Genes in the Intergenerational Perpetuation of Traits and Outcomes," professional associations have embraced social genomics and brought it to national and international scholarly audiences.[105]

Professional associations have also hosted panels on integrating genetic and social science research so that methodologies can cross-pollinate and become one. Bioscience associations like CHARGE have been the nurturing ground for social genomics consortia, but research has also been featured in panels at the American Society of Human Genetics, European Population Health Association, and more. Social science associations have held a number of methodological training sessions where sociogenomic approaches have been the pivotal basis for a new social science. Social science associations have also hosted cohort studies for presentations on their DNA resources.

In my visits to many of the latter form of proseminar, I have not only acquired material on how to access cohort studies' DNA but have gotten my hands on thumb drives with raw data. I have also heard a range of suggestions for ways to seize upon the strengths of particular datasets, such as which phenotypes to study (happiness, gambling, friendship) and which measures to triangulate (IQ, cognitive decline, gut biota). In recent years, talk of DNA has replaced talk of standard social survey measures in nearly all of the cohort study presentations that feed professional association contingents.

DNA-focused dissertations have also garnered the support of professional associations in the form of young investigator awards and dissertation awards. Dissertations like "Essays in Economics, Genetics, and Psychology"[106] and "Integration of Sociology with Genomics in Studies of Delinquency and Violence, and Social Stratification and Mobility"[107] have received the highest vote of support from associations like SSGAC and IGSS respectively, while also receiving awards from associations like the American Sociological Association and Trudeau Foundation of Social Science and Humanities.

Another way of gleaning disciplinary support is by looking at university support for teaching in the area. Several institutions have hosted social genomic courses. The Tinbergen Institute began offering "Genoeconomics" in 2015. As the course description notes:

> The course provides students with a structured overview of key concepts and methods in genoeconomics, discusses typical pitfalls, and outlines possibilities for applying genetic insights to classic questions in economics. It also emphasizes methodological issues in genetics research more generally, including study design, multiple testing, power analyses, calculation of effect sizes, accuracy of polygenic scores, and identification strategies to isolate causal effects.[108]

With sessions like "Heritability," which includes lectures on broad- versus narrow-sense heritability, using molecular genetic data to estimate heritability, and .GCTA methodology, the course draws on the expertise of a range of members of the social genomics community who work in a variety of countries.

Since 2015, the UNC Chapel Hill Sociology Department has also offered "Society, Human Behavior, and Genomics." As its description says:

> The course focuses on how molecular genetics can enrich the social sciences. Topics include a brief overview of genetics and how genetic and social factors combine to predict behavior. We also consider the ethical, legal, and social issues that sometimes complicate the use of genetic data to study human behavior.[109]

This course is offered to graduates and undergraduates alike, and is intended for students of all majors.

Oxford hosts "Sociology and Genetics" in its Human Sciences program which is taught to students from all colleges. Its description reads:

> This course will provide an overview of the new research field of sociology and genetics. We will focus on how social science can benefit from genetics and how the social environment interacts with and modify individual genetic differences. The course will provide the necessary theoretical tools but will also engage student on practical hands-on experience with software for analyzing genetic data. There are no pre-requisites for the course. A basic knowledge of genetics and statistics is helpful.[110]

It aims to train students in using genetic data and transform their pursuits into behavior genetics, gene-environment research.

And finally, Rutgers hosts the Summer Institute in Social Science Genomics, a workshop that "introduce[s] graduate students and beginning faculty in economics, sociology, psychology, statistics, genetics, and other disciplines to the methods of social-science genomics—the analysis of genomic data in social science research." As it states:

> The program will include interpretation and estimation of different concepts of heritability; the biology of genetic inheritance, gene expression, and epigenetics; design and analysis of genetic-association studies; analysis of gene-gene and gene-environment interactions; estimation and use of polygenic scores; as well as applications of genomic data in the social sciences.

This program focuses on training in software and quantitative methods but also on building mentorship relationships for future collaborations.

Though these are the only full courses being offered as I write, courses in a number of universities from Erasmus and Uppsala to Stanford and Harvard include curricula on social genomics. Like the "Political Psychology" course description from Exeter, which promises to "explore methods from psychoanalysis all the way to more recent approaches such as genetic analysis and neuroimaging," courses feature sociogenomics as the cutting edge and future of science.

These offerings and partnerships all point to the expanse of the social network of entities pushing for sociogenomic science broadly and social genomics specifically. As I've argued previously, the field which is autonomizing cannot be viewed without attention to these crucial pillars of support. At the same time, we must see autonomization as effervescently affecting not just science but governance, healthcare, and a litany of other public arenas. Even as initiation mechanisms like courses, certifications, and one-day degrees are set in place, the field, rather than siloing itself off from the world into a fixed domain of expertise, is part of a widespread collective becoming. Its interests align with the interests of a vast array of social bodies that are themselves shaped and co-produced in the process.[111] This support from all sides is indicative of the extent to which the sociogenomic paradigm is becoming a part of the many spheres of society, far afield from the immediate core of the science, deep into the very communities in which we live. It is also reflective of the growing alliance of a range of institutional interests in setting and meeting new standards of excellence in the postgenomic age.

CHAPTER 7

APPLIED SCIENCE

The team had been working with these miniscule variations in DNA for some time now. They had found success with a variety of mutations and were hot on the trail of a kind of sequence variation that could identify humans as unique individuals. Then one day they saw it: tiny tandem repeats in what everyone thought of as useless DNA surrounding our genes. They had a unique marker that could identify each and every one of us.

As Sir Alec Jeffries describes it:

> My life changed on Monday morning at 9:05 a.m., 10 September 1984. What emerged was the world's first genetic fingerprint. In science it is unusual to have such a "eureka" moment. We were getting extraordinarily variable patterns of DNA, including from our technician and her mother and father, as well as from nonhuman samples. My first reaction to the results was "this is too complicated," and then the penny dropped and I realised we had genetic fingerprinting.[1]

Immediately, Jeffries and his team began discussing applications. Paternity? Animal conservation? Forensics? To this list, his wife added immigration. Jeffries recalls: "That was when I realised this had a political dimension and that it could change the face of immigration disputes, especially where no documentary evidence existed."

The team's first case involved reuniting a young boy with his family. As Jeffries remembers, "It captured the public's sympathy and imagination. It was science helping an individual challenge authority. . . . The court allowed me to let the family know we had proved their case, and I shall never forget the look in the mother's eyes."[2] A year later and many familial applications on, the first criminal application was made. The case was a double rape and homicide of two young girls. After a DNA dragnet in which the males in the community where the crime had occurred had provided DNA samples, the criminal was found. Jeffries again heartened at the socially just applications his technology could provide.

Jeffries devised a special technology to assist in forensic cases, but his passion was with immigration and family reunification. He wanted to give recourse to families navigating a system set against them. Thirty-odd years later, the DNA fingerprint is known as a tool of surveillance. It is ubiquitously used in criminal investigations, including systematizing criminals into large national and supranational databases. And in 2015, a police program was established to triangulate computerized data from perps, including their criminal history, social media activity, and patterns in drug use, to predict who might commit future crimes.[3]

I share this story because it illustrates the point that there can be important unintended consequences for the science we create. Jeffries and his team rejoiced in inventing a technology that could help families win against a bully system. His hopes for the forensic uses of the technology were equally "with the people," helping community members and those invisible to the powers that be. But his technology has itself "mutated" and is now used in many ways that he couldn't have foreseen—and in some that he would likely not support.

Here I delve further into the public arena, yet this time I do so to characterize sociogenomic uptake in social institutions such as education, law, and criminal justice. I want to show what the consequences of this form of science are shaping up to be. I begin with scientists' own promises and prognostications for sociogenomic applications as quoted in major media reportage, widening my scope to experts working in an array of fields beyond social genomics. I then examine early adopters and the institutions in which they work to show the widespread demand that exists for behavior-predictive technologies. Finally, I compare these interests to social genomicists' desires for their technology. All in all, I reveal an unbridled optimism for genetic solutions in yet more realms of society, paired with widespread notions that society will be more just with sociogenomics as a part of it. This only demonstrates the growing ubiquity of the sociogenomic paradigm as well as the diverse sources of pressure for sociogenomic sciences to head in a genes-first direction.

Your DNA, Your Path

In the media, we see tons of predictions for where sociogenomic technology will take us. A perhaps all too obvious area that has continued to create a buzz is intelligence. In "A Genetic Code for Genius?" behavior geneticist Robert Plomin commends BGI's hunt for IQ genes, stating that the earlier educators can intervene on a student, the more likely that IQ gaps can be closed.[4] In "Success *Does* Depend on Your Parents' Intelligence," he is again quoted as saying:

> If we can read a kid's genome we can predict and prevent disease. If we can read their DNA, we can tailor the teaching to help a kid with learning difficulties. Surely it's worse to just sit in a classroom and sink, unable to read because no one has identified that you might have trouble.[5]

In "Is Being Good at Science a Genetic Trait?" neuroscientist Eva Krapohl says that "students are naturally inclined to seek out the types of experiences and education that works the best with their genetic propensities," so educators should be presented with information in different ways to enable them to choose what works best.[6] This would be a system based on "personalized learning" as opposed to "a one-size fits all traditional education," as collaborator Kaili Rimfeld calls it.[7] It would be a computer-based pedagogy that can be carried out at home or at school "in the way that makes the most sense to [students] without frustrating themselves by trying to learn in a way that feels more foreign," and that capitalizes on a student's personal genetic combination of "optimism, verbal abilities, or students' internalization of external issues."[8]

The promise of early intervention is also held out for at-risk youth. As developmental psychologist Dustin Albert and health policy analyst Daniel Belsky told Fox News, early intervention programs can mitigate a gene variant responsible for kids' responses to stress, which unchecked can lead to serious behavior problems.[9] In "Unlocking Crime Using Biological Keys," neurocriminologist Adrian Raine says that "we need to put the brain on trial as a prime suspect" and get youth on track before they reach the "path of violence."[10] Or as child psychologist Mark Dadds puts it in "Trouble, Age 9," "As the nuns used to say, 'Get them young enough, and they can change.'"[11] Researchers like Dadds want to help future psychotics exercise empathy, to grow the parts of their brains that make them less callous and unemotional. To this, sociologist Guang Guo adds that at-risk youth may benefit from having surrogate parents and possibly personalized drugs.[12]

Crime, as may be expected, is the most talked-about area of future application. In "Scientists to Seek Clues to Violence in Genome of Gunman in Newtown Conn.," molecular geneticist Arthur Beaudet recommends that genetic tests be used to determine sentencing in crime, particularly whether a criminal should go on parole or be considered at risk for recidivism.[13] Psychiatric epidemiologist Niklas Langstrom, who agrees with behavior geneticist William Davies in "identifying potentially violent offenders at an earlier stage"[14] and "identifying neurobiological pathways that might be amenable to treatment,"[15] adds that interventions can be targeted to families who are at risk due to shared genetic propensities for sexual offending.[16] As his collaborator Seena Fazel argues, families who already have a relative in the criminal justice system can be offered "relationship management and impulse control" advice.[17]

But scientists envision predictive testing around other problems that plague society, things like addiction, stress, and suicide. In the case of post-traumatic stress disorder (PTSD), scientists at UC San Diego who claim to have found PTSD-related gene variants are developing a test that can be used to target PTSD-prone soldiers for special programs and therapies.[18] Jill Harkavy-Friedman, vice president of research at the American Foundation for Suicide Prevention, endorses such a test for preventing suicide. In "Blood Test Could Predict Risk of Suicide," she says, "You wouldn't just look at one cholesterol level, you would look at diet, family history and what they ate that day," and rather assess their genetic tendencies and set up a program for increased monitoring.[19] Epigeneticist Zachary Kaminsky adds that predisposed soldiers could be asked to return their firearms upon return from war.

Indeed, drugs, diagnostics, and special therapeutic programs are recommended by a range of scientists and medical experts for more benign behaviors. One registered nurse told NPR that tests for oxytocin could be developed to aid in determining empathy:

> I just wonder, you know, if you could do a genetic test for where someone is with their oxytocin—much as you would do for serotonin or for dopamine, you know—there are medications that address those particular neurotransmitters. I mean, I don't know why you couldn't do it for oxytocin. In the end, I mean, I think, heck, you know, if I was feeling bad and I have to take a pill to feel better about my human beings, I would do it in a heartbeat.[20]

Similarly, in "Next Time You Go to the Doctor—Open Your Mouth, Say 'Ahh' and Take a Personality Test," neuroendocrinologist Kavita Vedhara says that it would be helpful to know whether a person was genetically predisposed to introversion, a trait associated with certain inflammatory responses:

> If you're confronted with a chronic condition you may well have underlying
> beliefs about your condition which influence how likely you are to engage with
> treatment, you might have an emotional response to that condition which might
> influence your underlying physiology and your ability to recover or to manage
> your disease, you may well have an orientation which makes you more or less
> likely to exercise.[21]

Scientists discuss targeted treatments for anxiety,[22] compulsive disorders like hoarding,[23] and overeating.[24] Scientists also suggest the potential for testing fetuses for inborn sexual and relationship traits, such as sexual orientation[25] and predisposition to anxiety.[26]

In reality, beyond the direct-to-consumer testing that I will discuss in the following chapter, sociogenomic science has almost exclusively been applied in legal settings. Hundreds of cases around the world have included a genetic defense.[27] The first case for which a court was asked to deliberate using social behavioral genetic evidence took place in 2002 in the United States, when Stephen Mobley was sentenced to death for murder. Mobley's defense team made a "genetic appeal" on his sentence, claiming that he had a defective MAOA gene.[28] The attorneys argued that "his genes had predisposed him to a life of crime," citing a 1993 Dutch study that associated a certain variant of MAOA with violence in one family known for generations of aggressive behavior.[29] Mobley's lawyers requested that the judge allow them to test their client for the gene variant, but the trial judge denied them the test. The court rejected Mobley's appeal because there was not enough scientific evidence to prove a causal link between the gene and violent behavior; however, they temporarily put a stay on his execution while other courts decided on similar matters. Mobley's appeal was eventually denied, and he was executed in 2005.[30]

In 2006, the MAOA defense found its first success. Attorneys representing Bradley Waldroup, an American man who hacked and shot his wife to death, brought forth expert testimony that he carried the low-activity form of the gene. The scientist who tested Waldroup, William Bernet, claimed to have collected the DNA of over thirty criminals. Though Waldroup was found to have told his children that he was going to kill their mother, the jury decided that having the MAOA variant meant that his actions could not be counted as premeditated. Thus they reduced his sentence from murder to voluntary man-slaughter and attempted second degree murder.[31]

In 2009, the MAOA defense was heard for the first time in a European court, and was also successful in reducing a murder sentence. The case involved Abdelmalek Bayout, an Algerian man who stabbed a Colombian man to death for insulting him on the eyeliner he wore for religious purposes. In his first trial, Judge Paolo Alessio Vernì accepted the defense's mental illness plea, com-muting Bayout's sentence to nine years and two months. In an appeal in which Bayout's lawyers introduced neuroscience and genetic evidence, including that pointing to MAOA, Judge Pier Valerio Reinotti shortened Bayout's sentence further to eight years.[32]

In 2012, Connecticut chief medical examiner H. Wayne Carver II ordered a full genetic analysis of Adam Lanza after the Newtown shooting in which he took the lives of twenty schoolchildren, six staff, and himself.[33] The Univer-sity of Connecticut Health Center joined the state department in the inquiry of Lanza's DNA immediately after. Forensic DNA specialists and molecular geneticists are still looking for associations between Lanza's DNA sequence and variants known to be associated with violent behavior and mental illness. Researchers report that they are targeting MAOA, autism, and schizophrenia associations.[34]

In 2014, Rene Patrick Bourassa Jr. was found guilty of bludgeoning to death Lillian Wilson in an Arkansas church, but with evidence of his low-activity MAOA variant his sentence was reduced from death to life without parole. Bourassa's lawyers employed the same geneticist that Waldroup's attorney's used, William Bernet, to argue that Bourassa's genetic code, coupled with his many traumatic life events, proved that he didn't have control over his actions.[35] Bernet was soon after quoted as saying that he believed that gene-environment research would be a growing challenge for the death penalty; however, that same year two other courts upheld the full sentence for convicted murder-ers who possessed the MAOA variant. One case concerned a triple homicide committed by Marcus Adams, an Los Angeles gang member who killed three people in an attempted car heist. A forensic psychologist explained that Adams had the MAOA gene and was at higher risk for violence especially with his history of child abuse and ADD.[36] The other case involved another car rob-bery and murder. Michael Anthony Tanzi was sentenced to death for abduct-ing, raping, and strangling Jamie Acosta to death in her automobile in Florida.

In an appeal, Tanzi's lawyers introduced genetic evidence indicating that he possessed an extra Y chromosome.[37] In these cases, the information was not enough to change ruling.

And in 2015, three death sentences were commuted when genetic evidence was introduced. In the *U.S. v. Barrett*, appeals judges reversed a death sentence for Eugene Barrett, who was convicted of felony murder and one count of intentionally killing a police officer. The defense proved that Barrett had genetic risk for mental health issues and that his previous defense had not shown enough evidence of his history and mental health conditions.[38] Similarly in *Chatman v. Walker*, Walker, who was found guilty of premeditated murder of two friends who had stolen drugs from him, was given a life sentence even though his charges of premeditation were not altered. Testimony from experts establishing his father's bipolar disease were critical to the appeal.[39] Finally, in *Delgado v. State of Florida*, Humberto Delgado, who was convicted of killing a police officer in an attempt to flee a robbery, won an appeal to reverse the death penalty. As with Walker, Delgado's charges were not reversed; only the sentence was reduced. Delgado's attorneys had introduced testimony to multiple mental illnesses, including PTSD and a battery of psychological diseases that they depicted as having genetic origins.[40]

The one educational arena in which tests have been applied is at the Chongqing genetics camp depicted in Chapter 1, where kids were tested for the gamut of IQ, EQ (emotional quotient), talent, and behavioral traits. As I'll discuss further in the following chapter, this program was carried out in partnership with the Chongqing Province school system but was not provided free to kids. Parents had to pay $880 to garner the full suite of services that the camp provided. Though there have been no further reports on the camp, scientists in China have made promises to continue to search for genes associated with these traits in order to provide diagnostics and therapies that can help schools tailor their programs to children's personal DNA.[41] These uses show sociogenomics' rapid development into public and private sector applications, which is ensuing with no holds barred.

An Interested Market

Given the newness of sociogenomics, I wanted to find out how prospective users felt about it. I asked a variety of people working in criminal justice, education, and law what they knew about the genetics of behavior and how it might affect their work now and into the future. I got a wide array of answers. While most experts express concerns about the ethicality of using the genetic research they have heard about and the kinds of tests possibly coming down the pike, nearly all want the opportunity to use tests. They see genetic information as an important weapon in their struggle to help members of society lead healthy and enriched lives.

In education, I heard interest from teachers and administrators who want to begin helping students earlier in their educational careers, either upon entry into school or even before. One private school teacher who primarily knew of sociogenomic science from the media, both in a fiction book she had read that centered on a family with "the murder gene" and from coverage of Adam Lanza's genomic autopsy, said:

> I just feel like I want to give every child equal opportunity, and I wouldn't want the way that I treat them or the way I approach them necessarily to be tainted by information. But at the same time, if there's helpful information, I guess I would want it.[42]

Her main reservation with research applications is the potential for branding kids negatively. She described the "handoff meeting" that takes place every year at her school, where the previous teacher of each kid passes on information about the student to the new teacher. She said that despite this tradition, they always want to start with a "clean slate" so as to "get to know the students and figure out the students out ourselves." She also wants to give the environment the benefit of the doubt, no matter what the genes say. Expressing a differential-susceptibility kind of rationale she said:

> I mean, there are so many things that affect behavior . . . they can thrive in a certain situation because of something, but in other situations it's kind of their downfall. But I do like the idea of it helping identify things that we can intervene in earlier, especially with something like a learning disability. If we knew that, [we] could start remedial work right away before it becomes a huge issue.[43]

Another educator, a charter school administrator who was familiar with sociogenomic science via the science literature more so than popular media, said:

> I would just like to see another strand of data. It's also like, what levels of data are you doing and who has the viewing capacity? I do think, in the hands of a really skilled and experienced administrator, that's probably pretty valuable

information that they would want to have, and that they could use to the school's and the student's benefit.[44]

This administrator's main issue with sociogenomic technology is that she wants to make sure that it will only be administrators who will deal with the information:

> I think that there's already a fairly high degree of confirmation bias very often in a lot of education, where people assume . . . a growth mind-set, meaning to assume that people can change and grow positively. . . . I think when I first heard that, I was like, why does that need to be stated? Because to some degree that's just the baseline that I operate from. But a lot of teachers over time no longer believe that, you know what I mean? So I think the kind of data that you are talking about is probably pretty valuable in the hands of a very skilled teacher or senior administrator, with the right level of judgment. But I think for a lot of people it'll just further their assumptions about good and bad kids. And it would probably negatively affect the educational environment that we would be able to offer for that child.[45]

She likened sociogenomic data on a child to personal health information and called such data a privacy issue as opposed to an ethical one. Having health information is good, but it shouldn't be used against a person. She gave an example of current practices to illustrate how genetic information could be misused if in the wrong hands:

> We are a very small pool of teachers already. When they know certain things about students and they start talking about it amongst each other, in irresponsible hands it becomes very much like, "Oh, student X is having a really hard time because of the situation at home. And that's why we don't really need to push them so hard on their academic work, because they are really, really upset." And you don't want that in a teacher. You want a teacher who is going to be, like, I need to make sure that I am holding you to high standards even in a compassionate but firm way, that respects what you are able to do.[46]

Again, the solution in her eyes is to limit the distribution of a child's personal information to top administrators and senior educators who have the training to deal with that information.

Another group of experts working with youth have another set of concerns. Juvenile justice authorities and probation officers are eager to have sociogenomic data on youth, but they are concerned with whether there will be suffi-cient treatment out there to help youth who "test positive" for innate behavioral disorders. As one probation officer summed:

> The term that I am most familiar with would be intergenerational incarcera-tion. It actually deals with a family, grandparents, parents, children, and their children, spending years in the system with one or more family members in the criminal justice system. I think if we dug into that deeper we would find that there are numerous or many symptoms along the way that are running through the families. Many families who are impoverished or lower income and even higher income don't wish to acknowledge that there may be something other than criminality involved when in fact there would be more on behavioral health or mental health issues that could be dealt with along the way.[47]

She gave an "emphatic yes" to tests or therapies to assist with identifying "be-haviors that are antisocial or criminogenic in nature, that would have a ten-dency to be of use to our system."[48] But she also expressed a common hope among people working in her field and with other youth services that there will be sufficient backup to help those identified as such. She is especially excited for teachers and early childhood educators to get ahold of tests so that, as she put it, kids won't end up in her office.

In adult criminal justice, there has been quite the buzz about sociogenom-ics. One state government official who had heard talk of tests for addiction, cognitive behavior, and predisposition to violence said that the probation of-ficers he works with are eager to have tests for these behaviors, but that the jury is still out on whether they will definitively tell us about something inherent in a particular offender versus portraying something environmentally passed down. He spoke of debates in criminal justice over whether measures for risk actually captured issues of need like poverty and untreated illness.

This official also recalled the XXY debates of yore, where criminal justice and law enforcement officials were alerted to the fact that "males that have those have a preponderance for violent behavior. . . . There are a number of characteristics that were kind of common—they were usually a bit taller and bigger than other men. And they all had this extra chromosome. And a dis-proportionate number of them were involved in the criminal justice system."[49] His concern was that with these findings "the technology just moves faster, the science in this case moves faster than our ability to understand and what would be a reasonable way to use it." Even though he believes that sociogenomic information about perpetrators can be "effective," he feels that the judicial system doesn't yet know enough "to utilize it without it being discriminatory, inflammatory, potentially biased and really just not helping for a just outcome in the case."[50]

Despite these reservations, in the courts attorneys see sociogenomic information as becoming increasingly important to due process and sound sentencing. As one attorney who works in a private firm who is familiar with the range of studies on social behavior told me, schizophrenia tests are already routine in legal proceedings. Another area where he has seen genetic claims is in sexual offending and "especially to sexually dangerous persons and questions of community-based probation." As he said:

> There is a large set of questions about how people convicted of sexual offenses should be treated, and generally the state has pretty broad latitude to impose all sorts of restrictions a lot of times in sentencing or probation matters dealing with sexually dangerous persons. All sorts of information will come into play and some of that will be, as I said, psychiatric information, which may partially draw on some of these studies of genetics and predisposition based on that. . . . I think one of the reasons it comes up so much with the sex offenders is because in part there is already an expert on hand to begin talking about the person's state of mind. And because there is usually a psychiatrist or a psychologist already involved in the case, a lot of that scientific literature can come in, whereas in a lot of criminal matters there is just never an expert witness.[51]

But for other behaviors that do not already require psychiatric evaluation, he believes that "if the information is available, defense attorneys will try and use it":

> The specter of eugenics looms large and sort of other sciences that cast a long shadow and raise equal protection issues, but a big part of it I think comes down to whether defense attorneys can use it, whether it's cheap enough essentially to get the testing done and to get an expert in to testify to it. And courtroom lawyers are pretty cautious to embrace change, but if legislatures say this is important, they can write rules to make that possible. I think it will come in probably from the defense side first, if it does, especially because the rules of evidence tend to be a little bit more—the evidence would have to be very probative. The test would have to be very clear in order for the prosecution to bring it in; otherwise it would be sort of unduly prejudicial.[52]

In other words, with trials definitiveness of results and cost are going to be of chief concern. In general, attorneys believe that with sentencing and probation there will be a much broader range of use, because these settings are "less strict" ("you can bring in almost anything you want in there"), and because "rehabilitation and propensity for certain actions" are key to probationary and parole matters.[53]

Of trial lawyers who have used "warrior gene" or "rage gene" tests, one indeed told me:

> They said, "That's hokey science. You're going to get embarrassed. He's going to get the death penalty because you're going to be bringing in quackery." But I didn't get that sense at all when I started studying science, when I talked to the experts, and when I evaluated this particular case. So I think there will be more and more of that. Probably hasn't been a lot of peer review on it yet to validate it, so it is kind of a new science. And a lot of people say that it's not valid, but I think it helps explain in the right case some bizarre behavior, and it certainly fed this case really well. So I think it will—it's become more accepted and studied.[54]

Like the educators I spoke with, he framed sociogenomics as a path to greater understanding:

> Nobody is going to say my genes made me do it, but if you might understand you do have this gene and you have the propensity for kind of a rage violence in a situation where it wouldn't be called for, I think it may open the door for us to understand some behavior that courts could consider and juries could consider in how to sentence people.[55]

A public defender who works in one of the largest cities in the United States said that though he has been successful in using neuroscientific defenses, he thinks that all mental health defenses are on the downswing:

> Have we integrated that into our work? Yes and no. I think that lawyers who are doing death penalty cases tend to rely more on those defenses in [this state]. The diminished capacity defense is no longer recognized unless you can show that the person is either unconscious or is acting involuntarily; it's very difficult to prove a mental health defense. You can introduce it to show lack of malice aforethought, which is the intent recorded for murder, but it's only a partial defense, and even if you show that there's no malice, you can still be found guilty of

second degree murder. So it's not used as commonly, but in death penalty cases where they are trying to save someone's life, it is more common and there are people who are experts in these areas.[56]

At the same time, he pointed out that the "unconscious defense" is making a comeback, perhaps paving the way for future uses of the argument that a person wasn't aware of their actions due to their genetics. He added:

> But everything is context, and that's why defense attorneys are more open to this kind of evidence, because we understand that it's all about context, unless you can present the context of a person's life.[57]

In other words, more information will equal more light shed onto causes and the merit of particular punishments.

Only one official whom I spoke with had words of caution. This chief of probation for a major metropolitan area said:

> What I know as a person who has been in law enforcement for more than thirty years—almost thirty-five years—is that the whole nature-versus-nurture discussion, it has not grown, in my experience, to be predictive. We have young people and adults who have come out of very loving, hardworking, highly ethical families who have gone down the wrong path, and we have handled it, and adults who come out of families that are in horrible circumstances. There has been abuse, neglect, violence, and they do very well. So I'm resistant to anything that assumes that we can predict someone's behavior based on their DNA. Now certainly I have boy and son who has Down syndrome. I certainly understand that as a result of his genetics he is going to have some challenges. And I certainly do not believe that he cannot have a quality, highly productive life. And so I am absolutely resistant to anything that would have us approach this work assuming that we can predict someone's behavior. I think we bring all of our resources to the table. We have some special tools that we use that are both static and dynamic to determine what will be most helpful. But even those we use with a very open mind about the person's potential and recognition that even within all of the tools that we have there still is an independent choice that the person has to bring to the table, that is more predictive of their success or failure than anything else.[58]

Again, this was the only less-than-eager position I heard. Other than the occasional link to past snafus in the history of criminology, experts in the broader public are sitting in waiting for tests and therapies to come down the pike.

What Social Genomics Thinks about Applications

Social genomics researchers have a different way of looking at the potentials of sociogenomic science. One place we can look for their vision is in supplementary material on studies, such as the FAQ of the "Common Genetic Variants Associated with Cognitive Performance" paper put out by the SSGAC in 2014,[59] which claimed that though practical applications to policy would be "premature," "studies like ours may provide scientists with tools that enable them to better understand the effects of government policies, which may eventually result in better-designed policies,"[60] help individuals project personal risks as in the case of "an individual planning retirement [who] may want to know about his or her risk of cognitive decline," and in the longer term produce "effective environmental interventions" that can reduce gaps and disparities. We can also consider debates at scientific meetings, where researchers have worried that the current child welfare system is tracking kids on the basis of environment-only data,[61] and that genotype will soon outrun social data as the most helpful predictor for policymakers.[62] Social genomics researchers have even penned sci-fi fantasies of a world where social genomics is dominant, as in one 2015 op-ed in *Nautilus*, which declared:

> The social world soon bent to this new auto-evolutionary reality: Not only did admissions testing for schools give way to genetic screening, the educational system fragmented into stratified niches based on specific combinations of genetically based traits: There were programs for those who were neurotypical and high on athletic ability, and others for those who were high on both motor skill and on the autism spectrum (a rather rarer category). There were jobs that required ADHD and those that shunned it. All in the name of greater economic efficiency.[63]

Across these public statements, we see a full disclosure that the sky's the limit for where sociogenomics can take us as a society.

In private, researchers have been simultaneously cagier and bolder, starting out with caveats about the limitations of the current science, but ending up outlining a future where those limitations have been surpassed. One political scientist said:

> Right now there are a lot of fears about people using this data for prediction. I think those fears are exaggerated just because the data is useless for prediction right now. If it were the case that—when they become useful, and we expect that to happen in intention, then I think there will be interesting ethical debates about what the boundaries are.[64]

A sociologist agreed:

> I like having more knowledge, but this is very small pieces of information for
> any one person just because this is that classic bias of taking so much informa-
> tion that's kind of for the general public. It's just broad information. It's not for
> any particular individual. And applying it just doesn't make a lot of sense at
> this point. Same goes with longitudinal data; I mean, I guess if you were to get
> your data and track its decline over time, maybe you could say, "I'm aging," it's
> moving faster than you might anticipate, "What am I doing wrong?" But I don't
> know if it's worth it at this time for a specific person.[65]

An economist put it this way:

> I think that there are a lot of useful policy implications that have to do with just
> an individual knowing their personal risk. I think it's something that could be
> really valuable. I don't know the value of larger policy changes based on geno-
> type. I am more skeptical they could be generally valuable. I could see us some-
> day being in a world where we know more, but like I said, I think the potential is
> there, but I feel like we do need to start thinking about the ethical questions . . .
> if it turns out that we do find something big. But I think that they are not large
> enough that we need to be immediately worried.[66]

Researchers downplayed the need for worry now, but noted the coming tsunami of personal social genomic data.

The thing is, nearly everyone I spoke with seemed to believe that the tsunami is already headed our way. Some researchers cite the varying landscapes
of scientific production in different countries and regions of the world, where
people are already pursuing preimplantation genetic diagnosis, sex selection,
and gene editing. One researcher, who said that social genomic data could "really help planning your life" and that even if "nothing's one hundred percent
predictive, having information is better than not having information," told me
about a Scandinavian member of BGI's Cognitive Genomics branch who was
working toward a world structured around inborn hierarchies:

> And he's like, "You know, it'll be great when we can have the janitors just be jani-
> tors." And he didn't say that, but he was like, "I want to have my kid—I would
> like to have a bunch of embryos fertilizing and to take the best one and implant
> it." And I was like, first of all, that technology is a long way off because even if
> we have a great—I think too now, we could tell you which kid has the likeliest

highest IQ potential. For example, by running genome-wide stuff and creating a cell score, and so we could do that. But you have to take a cell out of an embryo. I don't know how much you're going to damage an early embryo by taking one of its cells . . . at, like, the thirty-two-cell stage.[67]

This researcher's point of contention with his colleague's aims aren't so much about the research application but the viability of the technology.

I heard further on the topic from others:

Height? Certainly. There's high predictive power now with people with that kind of data. Schizophrenia? They can predict twenty-five percent of the variance. So those are things I would think people would want to screen on. But my understanding of PGD, pre-gestational [diagnosis], is that they can essentially look for chromosomal abnormalities, but they don't have the resolution yet, or at least clinically. Maybe some lab somewhere's doing that research, but it hasn't diffused clinically to be able to actually look at SNPs and sequence them and say, "Okay. Here's just predictive height. Here's your predicted schizophrenia risk. Here's your predicted IQ."[68]

This researcher, who is affiliated with the SSGAC and its work with the personal genomics company 23andMe, forecasts that with their greater predictive power for educational attainment and cognitive performance, test makers like 23andMe will likely eventually be able to offer people "predicted intelligence" and "predicted education."[69]

Researchers also stress the already unequal systems of tracking in place today. As one sociologist attested:

What matters is the combination of genetics, environments, sometimes working additively and sometimes working interactively. So I simply don't see us coming to a point where we're going to be able to identify a specific gene and say, "Okay this person needs to go to this school," or we're going to identify a system of genes and say, "Okay this person is destined to become a criminal." But if we sort of take away the genetic part, we sort of do this anyways. I mean, we look at students and we test them, right? And based on the results of those tests, they might go into college prep courses or advanced courses or basic level courses or whatever it might be. If you believe the findings that are coming out of the behavioral genetic research, really, these are just manifestations of genetic effects. So, if we think that test scores and IQ and those things are heritable, then these test scores that we're using to sort of place individuals in different paths are really just manifestations of genes.[70]

An economist averred:

Any kind of difference between people and especially differences that might be correlated with performance could be used as a basis for tracking or separating people into different schools, but it doesn't mean we should do it, and I think that we could already—you could imagine giving people IQ tests at age five and then tracking them based on that. In fact there was something like that in my own school; there was like an enrichment program that we all took IQ tests when we entered the public school and some people got sent to the enrichment program at a certain time of day.[71]

Like many others, these researchers ended by saying that currently we don't have enough people working on gene-based tracking, and so it will likely remain unpopular until the science moves forward. Nevertheless, they have full faith that it will come to fruition soon enough.

Given this faith in the science, how exactly do social genomics researchers see the world changing? They do so in terms of better treatment for individuals and seismic social shifts as a result of gene-environment-based policy. Take, for example, education. Researchers talk about "targeting educational resources to kids that might have a kind of deficit,"[72] diagnosing learning disabilities, and characterizing "children's health and well-being as they enter into schools—a genetic sort of portfolio about them that gives them resources and risks."[73] In fact, a genetic profile for kids was on the minds of many, as this political scientist's remarks detail: "So if you have a certain social background, a certain upbringing, and a certain genetic profile, and all those things interacting, you have a certain learning style that maybe we might tailor education for . . . very much the way they are trying to do personalized medicine."[74]

This researcher gave the example of SSRIs (antidepressants) and the standardization of treatment and then said:

Instead of that trial and error, we might be able to learn some information about somebody's neurobiology, somebody's genetic profile, and say, "Oh, well, this regimen has a higher probability of working, so we will try that first." And that's really the only thing we will get from these is understanding systems and probabilities, but probabilities aren't guarantees; it's just [that] children who were abused have a higher probability of becoming abusive. It doesn't mean that we damn them to being abusers.[75]

Describing what many others articulated, he said treating social behavior will become like treating cancer: "You have a higher probability, so here are some things that we could use to inoculate you; let's lower your risk."[76]

Researchers see using the genetic profile as a process involving top-down as well as bottom-up changes. The political scientist just quoted talked about giving kids with innate aggression susceptibilities genetically tailored educational environments to "stop the environment from aggressing against them."[77] Another researcher said that once we know more about traits that facilitate academic success, like "conscientiousness and perseverance," as well as those that prohibit it, like those responsible for learning disabilities, we will be able to "help people flourish" and eradicate contexts that "stunt people."[78] Yet many also spoke of empowering parents to request nourishing environments for their kids. As one sociologist argued:

> So the example I have been using is there is a GWAS on ADHD; parents could have their kid genotyped to decide based on the kid's genotype and the GWAS to show a high likelihood adult in ADHD. They can perhaps request an intervention be performed with their kid in their schools before the kid was even showing symptoms.[79]

This researcher feels that it's imperative that social genomics be studied with perseverance so that schools can get to the point where these requests can be addressed.

When it comes to adults, social genomics researchers also envision top-down and bottom-up interventions. For example, with crime, researchers see the eventual creation of interventions that prevent potential criminals from ever committing a crime. Take this political scientist, who studies terrorism:

> One of the things that we have been curious about is if we can better understand why people engage in extreme political acts, up to and including political violence; maybe we can figure out a better way to head that off. And especially in today's world, that's a pretty big pressing issue, when people are strapping bombs to themselves and blowing themselves up and killing others in the name of some underlying ideological or theological belief.[80]

And regarding prisoners, this sociologist of delinquency said that the state could create better environments for incarcerated expecting mothers who have a propensity toward depression, thus giving birth to troubled babies:

> Thinking about the prenatal effects and how they influence early child development, and then immediately after that talking about interventions that have

been created within child development and psychology which pretty much counter those effects. If that's adopted to, say, a mother who is experiencing parental incarceration, or it's an underlying factor that's associated with some of the related issues with family instability or economic instability—depression and things like that—it's policies that can help to reverse those effects.[81]

But with personal financial and health decisions, researchers envision the creation of incentives for individuals to make better decisions for themselves. As one economist said:

I can imagine at some point it might be the case that we find that certain individuals are responsive to sort of financial incentives in a very profound way and certain people aren't, and there might be some genetic underpinning to that, and so genetic testing could help. If we're looking at addiction recovery, there are things we are intervening and trying to help people with whereby they'll help you live better lives, to direct certain types of treatment at certain types of people based on whether they will respond to financial incentives or not, or whether they will respond to a certain type of treatment or not.[82]

Similarly, a health economist talked about incentivizing better personal healthcare for those individuals who don't use health services, take their medications, or visit their doctors:

I see people having a certain genetic risk score and then either seeking out or not seeking out health services. . . . Do they get any routine check-ups? How often do they go to the ER? How many hospitalizations have they had? These are kinds of both preference measures, regarding beliefs in healthcare, and effectiveness of healthcare and the value of healthcare, as well as actual kinds of reported behavior in terms of it . . . People who say, "Yeah, there are some things that I could do to improve my health kind of on my own and improving my health is kind of in my own hands," that seems to be partly genetically influenced. We found that use of prescription drugs as well is highly genetically influenced.[83]

In other words, individuals could be empowered by their genetic profile to do more to improve their use of health services and to be compliant in their healthcare.

Researchers envision broad societal changes in the near future because of what's already going on in the public. As one sociologist said:

I could see certainly on dating sites, people linking their OkCupid or whatever profile to their 23andMe data, and scores are calculated (and that doesn't need to wait for the scores to be predicted well). And then I could see that, right now, pre-gestational genetic testing, I mean prenatal genetic testing, is not when they are deciding what embryo to implant; right now they can do full chromosome scans, where they can see if there are major duplications or deletions of chromosomes, but they can't amplify enough DNA reliably to like do genotyping assay or full genome sequencing. I think that when that's changed (which I don't know why that would be an enormous problem; I mean, I understand that you're taking one cell, and you're taking the DNA from one cell, and you're trying to amplify it, and there are lot of impurities and so it's very hard to maybe factor that in the lab), it seems that will be calculated at some point and then we will have a mechanical world where each potential embryo is assessed and you get the scores on a variety of dimensions and where you make the decision on which to implant.[84]

Researchers see dating and mating transforming society and the family entirely. As comments from one political scientist further illustrates:

It could be like I avoid marrying you because the combination of you and me is potentially bad. Or maybe we don't go that way and we end up checking at the genome of our future child. . . . And then people are going to try to make changes based on having a smarter child or better-looking child or whatever the trait they think about, because just down the road is the ability to edit the genome. That's going to lead people to make these choices about things that are probably morally questionable whether or not people should be able to do that.[85]

This researcher, like most, thought that designing babies to be superior was the worst-case scenario, but a very real possibility. When I asked if we were really headed that way in America given the country's reluctance to provide universal healthcare, he replied:

No, but other countries are investing more. And even like New York State was talk-ing about every newborn—now they take blood to do a couple of tests. But I mean you could have all of the information possible get to parents, give them guidance on a couple of key things, but then give them all of the genomic information of their newborn child. And then they could go somewhere and have that interpreted for them, or maybe they could even try to do it themselves. That could potentially lead to problems, but that's a discussion that folks in New York have had.[86]

These comments show that eugenics is not a taboo subject or vision for so-ciogenomics by the very scientists who produce it. It's more of a reality than many would want to accept.

Social Genomics' Vision of Social Justice

The bold new future that would have genetics direct us to a more utopian society is a common view among the producers of sociogenomic science. In fact, I encountered many arguments for why it would be *more* rather than *less* responsible to allow sociogenomics to unfold. In a variety of ways, akin to the genome scientists I wrote about in *Race Decoded* who were certain that they were fighting the good fight, sociogenomic scientists see sociogenomics as social justice.

One kind of argument out there is the idea that sociogenomics will bring about "early warnings." One researcher illustrated:

> You get your baby genotyped and you find out that there is a decent chance that they are going to be dyslexic, and you put them in early intervention for dyslexia. And my guess is that there are going to be things that we can do, where if we focus our resources on people who are at higher risk, it will actually be affordable to develop early intervention for some of these characteristics that are influenced by genes.[87]

This scientist, like others I spoke with, said he couldn't imagine that kind of educational tracking being any worse than it already is given the classism inherent in most current public educational systems. He predicted that the area where it's "going to be really useful to have genomics to play a role in is in terms of social policy," where it can show "that for half of the society, that we need to be concerned about helping, that it's not genetics, it's not genetics at all." He continued:

> For the above-income people, genes are everything. For the below-income, the environment is everything. Why? Because all you need is enough to eat and a roof for your head and reasonable safety, and then the only thing that's left is going to be natural ability. But if you are a really smart kid and you are born in a neighborhood where you are not safe—people are firing guns into your house or something like that—it's going to be hard for you to develop to your full potential.[88]

His words sum up the prevailing sentiment that policy and public administration would do better to incorporate sociogenomics.

Better policy was a common thread among scientists who agreed that sociogenomics would be the best litmus test. As one put it:

Things like, well, the talk about MAO. I think the Italian appellate court in Tri-
este was the first to do it, and then we saw it kind of bleed over, and now it's in
the state courts that you want to reduce sentences, or absolve some level of guilt
for people who commit certain crimes based on certain genotypes. If anything,
our research has shown that's just crazy. It would be the same as saying, well,
it's a prison group in a bad neighborhood. We should just give them a lesser
sentence.[89]

This researcher talked about agency and the "tiny tiny effects" of many genes,
claiming that sociogenomics could "prevent bad policies too."

But an even more common view is that sociogenomics can prevent inequality from the get-go. Take IQ. Many researchers cite the statistics that show
that higher IQ leads to better life outcomes. As Plomin argues: "Little genetic
differences become bigger and bigger as you go through life. Bright kids read
more, they hang out with kids who read more."[90] Or as Hsu more ominously
warns, since cognitive enhancements are the future we need to go forward with
a framework for bringing equal access to all, "the alternative [being] inequality
of a kind never before experienced."[91] The idea here is that sociogenomics is
inevitable, and that we need to make sure it is used for the good of all.

Indeed, like other sociogenomic scientists working in the genome era, some
social genomicists maintain that we are better off with them leading the way.
As one said:

I think the reality is that a lot of this information will be used by corporations
for purposes of profit, in ways that people are unaware of, and may or may not
approve of if they were aware of it. And that's going to happen regardless of what
scholars do. I mean, it's like big pharmaceutical companies; they are just going
to go their own way, because there is money to be had. . . . But I also think that
it's really, really important that scholars continue to do this work, because they
are the only balance against the corporations that are going to be doing this for
much more individual profitable reasons.[92]

The idea here is that this group of researchers isn't in it for profit, so they are
more trustworthy. And after all, they do know the intricacies of what the science
is telling us better than anyone who might just slide in to make a quick buck.

But are the scientists responsible for this field of research really not about
profit? Most researchers I spoke with said that they hadn't yet partnered with
biotech firms to make tests. Many bristled at the idea. But they also thought

it was only a matter of time before the field would have to grab the reins of the market and do research and development for applications. Remarks like "I definitely receive more than enough junk mail, is how I would characterize it, so whether it's on LinkedIn or just sort of garden-variety spam, it's targeted; like they have my name, they know my position, they know whatever paper they are referring to"[93] were followed by "It's the funding organization that has asked me if I am going to do anything more [with] the results and something like a web application or an app or something,"[94] indicating the pressure from partners to produce for the market.

I heard from a number of people that pharmaceutical agencies were already aggressively seeking to make drugs and diagnostics. As one researcher put it:

> I've had e-mails from businesspeople in California when the leadership paper came out, so the leadership paper won best paper eventually the year after. This was about a SNP RS4950 leadership development [gene] showing leadership. So they came up saying, "This is revolutionary. We would like to build business around this."[95]

Another said, "We had a couple of people contact us who were in various stages of start-ups where they were interested in using this as a diagnostic tool for whether or not two people should be friends."[96] This researcher said he "suspected these kinds of tests are going to be developed at some point over the next twenty years, and they will be more powerful than they are now."[97] Yet another mentioned having had "doctors ask me about when do I think something will be ready. And certainly the army would be interested at some point in having scores."[98] And then there were still others who had been paid or been offered to be paid, as one person put it, "an exorbitant amount of money,"[99] to consult on marketing and even political and military campaigns. Social genomics researchers are clear that they don't want to be beholden to this profit motive, but most also see a truly ethical outcome. As one researcher who works on projects with 23andMe asked me: "So just think about it. How would you react if you would be a 23andMe customer, you're nineteen years old, and you see that the genetic chance of going to college is way below average? Would that motivate you or would it demotivate you?"

When I said, "Motivate me," he said, "I think that's probably what would happen to a lot of people." He, like so many others I spoke with, feels optimistic about people's future responses to their own DNA.

This kind of vision for the future may one day lead to a world in which sociogenomics rules. After all, sociogenomic entities are already partnering with 23andMe, the UK Biobank, and Decode. But what of the unintended consequences that may come to prevail?[100] Will applications simply motivate people? Or will they become embedded in institutions rife with inequality, ones that are dedicated not only to helping but to surveillance? Or along less ominous lines, what if to save time or resources strapped institutions use applications to sort people? A *GATTACA*-esque future could be in the cards regardless of how conscious experts are of their goals and objectives concerning the use of applications.

No matter what's in store, we must move beyond relying on expert intentions or awareness to ask the bigger questions, "Expert for whom, and to whose benefit?"[101] Experts with the best of intentions, like the individuals I spoke with here and also the scientists I write about throughout the book, can only see so far ahead (and so far behind, in terms of the historical implications of their work). They are beholden to the charge of their positions and their sponsors. No one person will have the whole picture in view. Something as simple as a genetic score on a person's aggression levels or a prediction of their likelihood to do a particular job well can be used differently by the range of social institutions that they move through in their lives—education, health and medicine, work, and more.

There's also the looming concern of stereotype threat that I mentioned in the introduction to this book. Stereotype threat is like a veil that falls over a person's eyes and has them see their potential in terms of the stereotypes that seem to apply to them. The threat of performing badly (or well) leads that individual to live up to the stereotype. With evidence of a person's innate abilities, how will sociogenomically characterized people feel about themselves? How will they perform? Better? Worse? The self-fulfilling prophecy that sociologist Robert Merton theorized nearly a century ago is more relevant than ever with sociogenomics on the scene, as is the law of unintended consequences.[102] Therefore, we must heed the potentially devastating implications of the sociogenomic paradigm before irreparable damage is done.

CHAPTER 8

THE BUSINESS OF SOCIOGENOMICS

I leaned against the AV table in the middle of the auditorium and placed my coffee cup on the unoccupied seat in front of me. The experts on the stage were just finishing up a discussion about eugenics. Classes had been brought in from all around the New York area, and over the last few days students had joined the conversation, becoming vocal about their views, staking claim to positions. Pretty much everyone there was in agreement that eugenic beliefs, and the idea that some traits should be encouraged in the gene pool while others should be eliminated, was wrong. Even the youngest of the middle-schoolers showed a sophisticated awareness of the malleability of traits like intelligence and physical strength.

The discussion moved toward the issue of inborn drives. The man holding the mic addressed the kids in the audience. Do we have certain inescapable drives? Is there something about us that's genetically programmed to, say, fight or flee, to cooperate or to put self-preservation above all? Without skipping a beat, a number of kids chimed in. Humans *do* have built-in cooperation mechanisms *and* warring mechanisms. And with these, we will always naturally put our family first, our biological relatives. One kid raised the point that there may be nonbiological kin and community defenses, but overall the youth present agreed that there are genetic behaviors controlling us at every turn.

Where do these kids get their ideas about our inherent nature? How can they be so sure? In informal conversation, I gathered that they surmise all this from conversations with other kids, listening to their parents and other adults around them talk, and also from movies and TV. When I later spoke about this with other young folks in San Francisco, I found that many are privy to the gist of what the news media has been saying about behavior. And a lot of their conversations and the sci-fi programs they discuss are themselves riffing off of news coverage and the vast world of testing on genetics and behavior. The booming business of "genes found" has been making its mark on the minds of the future generation.

Indeed, the media is increasingly casting a spotlight on genetic research on behavior, and has therefore become a fundamental source of notoriety for the science and for the specific brand of gene-environment research that portrays genetic causes at the expense of social factors. From 2005 to 2010, articles ramped up their coverage of genetic causes of behavior. And from 2013 to 2014 alone, the number of articles covering sociogenomic topics increased nearly tenfold. As with prior media coverage of the genome era, today's media has propelled sociogenomics without due diligence on reporting whether study findings have been validated, whether they are generalizable, and whether there is debate around them. The media therefore serves as a conduit for the haphazard geneticization of phenomena vastly distant from biology and the body, and a proponent for science that also takes those shortcuts. It literally creates value for the sociogenomic paradigm, popularizing a genes-first framework as the new truth of human being.

But the media isn't the only one selling sociogenomics (or the genes-first lens that encourages sociogenomic fields to continue to molecularize and be all about genes). The market is, too. Character tests, talent tests, and various personal genomics services put sociogenomic frameworks in the hands of everyday people. While these commercial endeavors are being undertaken by scientists and businesspeople outside the field of social genomics, often far afield from gene-environment research, their uptake encourages unsubstantiated notions that all of the genes that gene-environment researchers study hold predictive value for people's futures.

You've probably heard the saying, "There's an app for that!" when you've wished aloud for some mechanism to organize some part of your life or improve your day-to-day systems. Soon that catchphrase may be replaced by "There's a test for that!" While there are still just several dozen sociogenomic applications in circulation, more emerge every year and they come from new markets that target new social phenomena. It is an understatement to say that this industry is growing.

Here I provide a snapshot of the commercial realm at the start of the new millennium. I also compare the use of sociogenomic biotechnologies in various countries, graphing where technologies have taken off and where they have met with resistance. I show that the market in technologies has generally developed unchecked in the United States and abroad, and this despite the fact that many tests market and normalize eugenic strategies that could have disastrous consequences for society.

Inborn Talent

Just when you thought that talent was a product of hard work and practice, a number of personal genomics companies have emerged selling tests to determine whether your child has inborn talent. Talent tests are marketed to parents who might be curious about their child's future. Ads are also loaded with social norms about how to best parent and how to capitalize on a child's inborn resources for the good of society.

Take Smart DNA Testing's Children's DNA Discovery test. Smart DNA Testing, an Israeli company that sells tests worldwide, invites parents to "empower your child's well-being and future. . . . With just one easy test, you can empower your kids from the start and receive lifetime access to their personal DNA report."[1] Like other companies who have cornered the market on personal genomics for adults, Smart DNA Testing tests for a combination of traits including physiological traits such as near-sightedness and height. But the bulk of traits it examines in its tests for children are behavioral. From quasi-physiological traits such as athletic potential and memory performance to patently social traits such as risk taking, avoidance of errors, and learning patterns, Smart DNA Testing makes predictions on a child's future way of thinking and acting in the world. There is even an analysis of something it calls social support, or how a child responds to crisis: "Does your child actively seek social support during times of stress? Does he prefer to handle things on his own?" These analyses all feed the claim that a genomic test can help parents unlock their child's potential "from the start."

Smart DNA Testing is unique in its marketing in that it starts from the premise that parents will suffer without the information that the test provides. Its website asserts, "Being a parent is one of the most difficult jobs on the planet." The company offers the Children's DNA Discovery test to help overburdened parents cut to the chase with training, education, and even the selection of a child's activities. Smart DNA Testing also aims to alleviate parental stress and rechart parenting strategies toward more positive outcomes. As it says in the case of "avoidance of errors": "Does your child have a hard time learning from his mistakes? A Smart DNA Test could reveal if his genes are to blame." Similarly, with the trait it calls morning/night person, it proffers DNA as a cause of "night owl" and "morning person" behavior.[2] Implied here is the idea that parenting is easier and more efficacious when tailored to a child's DNA.

Companies like Smart DNA Testing that offer talent tests to parents also offer similar tests for adults. Another Israeli company, the Makings of Me, is a case in point. The Makings of Me offers the test My Child's DNA Insights alongside the My DNA Insights test, which examines all the same alleles minus the alopecia (hair loss) allele for male adults (Smart DNA Testing works pretty much the same). While to adults it markets traits in the present as a "response to a certain diet" or "excelling in endurance sports," to parents it markets traits as future realities like "overweight potential" and "athletic potential." And again, with what it celebrates as "an affordable test" that can "unlock the secrets of how your children's genes influence who they are and what they can become," it instructs customers to be better parents by making a parenting program tailored to a child's DNA.[3] The orientation of these companies is thus prospective. Get with the program now, so that a lifetime of better mind and body may unfold.

The Makings of Me states clearly that it is all about "the soft and the fun sides of genetics knowledge and applications, which may help our customers to improve their life style and their recreational habits." But some companies aren't afraid to send a more ominous message. Oogene, a Chinese company, asks parents, "Have you found out your child's nature?" which it follows up with a litany of warnings about needing to know early enough to develop a child's talents. Will you fail to make full use of your child's early childhood neuroplasticity? Will you never develop your child's latent abilities? Worse yet, will you make your child insecure, with low self-esteem, because you haven't let him or her "flow" in the right stream and thus haven't provided sufficient opportunities to "win"? Oogene also plays on parental image, if not insecurities, claiming that a parent's cool factor will go up by using its genetic test to parent.

Oogene claims to test for far more behavioral traits than most other companies—a whopping forty-seven behavioral traits that it groups under IQ, EQ (again, emotional intelligence), and artistic, dancing, musical, and sports potential. But it is similar to other companies in the wide latitude of its claims. Next to a picture of a girl passed out from boredom on a stack of books, it writes:

> Stop wasting your money on your child by signing up for multitudes of enrichment classes! Parents used to and still sign up for multitudes of enrichment programs, inundate their children with tuition, line up classes upon classes of drama lessons. What for? To give their children an edge. . . . Is it really necessary to waste so much money? Hefty sums of money invested in enrichment classes may be avoided with a clear direction of which choices to make.

It makes personal sociogenomics a money issue, selling it as an economic boon for strapped parents.

Another Chinese company, Genetics Center, similarly warns parents against setting children in a "wrong direction," leading to "money, time, and efforts wasted." It then goes on to frame nongenomic (aka "bad") parenting as setting kids up for a lifetime of failure. It states:

> Ultimately, it is difficult to compete with another child who has advantages of talent potential. So check the child's genes, as knowing the child's innate talents and traits will be an additional advantage. Parents can plan the child's path in academic, vocational and social competitiveness based on their potentials, and maximize their inherent abilities and avoid entering into a field which appears to be suitable but in fact will not bring any achievement or success.[4]

While this framing may not be as pointedly eugenic as the negative and positive antinatal and pronatal policies that include forced sterilization and racialized immigration, it is what sociologist Troy Duster has called a "backdoor" for eugenics. It encourages consumers to chart their future and direct their families by a person's so-called inherent, genetically determined abilities.[5]

The American- and Chinese-based company Map My Gene's Inborn Talent test is even more explicit in this regard. It asserts a vision of "improving the overall standard of living for the people of the communities where we operate in," and it claims that its mission is to "transform the landscape of health and educational institutions" by DNA-based social interaction.[6] Map My Gene refers to parenting by DNA, what it calls "Genetic Profiling-Planning," as the only correct "develop programs" for youth, and it even offers specific career and training paths for clients. In other words, it builds its product as the only basis for a healthy society. In its battle cry, "Attention to All Parents, Educational and Medical Institutes," it carves out a place for sociogenomics to lead citizens to a superior social order. Whether by "inherent intelligence, personality characteristics, superior athletic performance, behavioral problems, social/entrepreneurial skills, musical, linguistic, performing or dancing ability, and other areas where he or she can excel," Map My Gene offers their Inborn Talent test to promote "True Optimal Wellness" for the individual and society, specifically the communities which Map My Gene serves.[7]

Even if test makers didn't make such grandiose proclamations, there is an insidious way that the tests they promote link up with more widespread eugenic beliefs, such as the racist fallacy promoted by *The Bell Curve* that Asians and whites are genetically superior in intelligence to everyone else on the planet.[8] Talent test makers promote the idea that some groups are genetically gifted and destined for intellectual success while others are not, by drawing on racialized studies of behavior. Some companies merely provide links to the studies that analyze things like IQ and athletic ability by race. Others arm their clients with ethical defenses. Map My Gene, for one, offers a link to famed geneticist James Watson's op-ed on racial IQ, "To Question Genetic Intelligence Is Not Racism."[9] This article lets parents off the hook, stating that DNA sequencing will show that genetics, rather than environmental factors, can predict and explain behavioral problems. Sentiments like these are particularly dangerous because they reinforce the notion that there are natural hierarchies between races and civilizations. And no matter where a company is based, there is a trend among talent test makers to market specifically to the Far East and the West. The Makings of Me, for example, has all of its website available in English and Japanese. Map My Gene offers English and Chinese brochures. Tests are being devised and sold under the auspices of revealing innate truths in the only two places in the world that benefit from the perpetuation of eugenic ideas. Therefore, there are no checks on how far these ideas might go.

Another way that talent tests advance eugenic beliefs is in promoting the notion that cognitive function is programmed by one's DNA. As shown, test makers openly offer IQ DNA tests, thereby giving traction to the idea that IQ and intellectual aptitude is in our DNA. But what most don't realize is that talent test makers also recode as genetic a litany of neurological processes that many people take for granted as deeply biological. The Learning from Mistakes test provided by Gene Planet, a Slovenian company, exemplifies this:

> Numerous times we have made the wrong decision and its consequences were unfavourable. But the cause does not lie only in our thinking. A mutation in the DRD2 gene can also be responsible, because it can cause a smaller number of

dopamine receptors. They are responsible for remembering our wrong choices, which in turn enables us to make better decision when we encounter a similar situation.[10]

Its Episodic Memory test rings the same: "If you very quickly forget what you had for breakfast in the morning or you have difficulty remembering dates, the reason may be in the KIBRA gene. Individuals with certain gene variants require more brain activity to recall certain information."[11] The language here is clear: one gene may be responsible for an entire sector of an individual's neurological system. And since the range of behavioral processes that have an impact on our moment-to-moment lives are controlled by genes, it is imperative that tests be taken by every responsible parent. The problem with this spiel is that no responsible scientist would reduce these behaviors to a single gene.

Tests may encourage a eugenic mentality and a new rash of practices as well. When 23andMe offered testing on alleles associated with intelligence (a feature reneged in 2013), an enthusiastic buzz formed around IQ and cognitive function among its consumer base. The buzz began when one of the company's own bloggers prompted clients to check their so-called intelligence genotypes against the scientific record.[12] A bunch of people then began crowdsourcing information on IQ genes.

People wanted to know who had "2 copies of the gene for intelligence." Parents reported their kids' genotypes, asking why some were "exceptional" and others were "dumber."[13] Others shared the stats on all of their family members. Still others shared diploid status plus test scores like SAT and IQ. Within months, EQ was thrown into the mix, as well as happiness genes and genes for kindness. 23andMe simply reminded test takers that "while some aspects of intelligence are . . . difficult to quantify . . . studies estimate that in early childhood about 25–40% of individual variation in measurable intelligence can be attributed to genetics. In adults, this number increases to about 80%."[14]

Race reared its head in the ensuing community bulletins, in test takers' pursuit of who was most genetically endowed. One asked, "Should we start marriage between people who have 2 copies of the intelligence gene?" Though a couple of people responded negatively, most were supportive, motivating this test taker to declare:

> Many people with out knowing are had made eugenics, when white man chose
> to married a white woman and he prefer not married other ethnicity it is mak-
> ing and am sure that many of you had married a person from you same ethnicity

and even some of you would like to keep some of you psychical features in the lineage of your family. I don't care about races at all, I prefer care about my son or daughter get the best genes that will help for their future, and am sure that soon this idea will be more popular, and we may end having sons with less risk to have cancer, that live longer, with increases points in many of the beneficial genes, with low high blood pressure, high IQ, and why not also good for sports, etc., etc. Or you don't wish the best for your future sons? They would have more probably to live longer, to be healthy, smarter and happiest.[15]

Though the claim here is that race is not the issue, this person makes clear that eugenics should be okay because people already assortively mate on the basis of race.

Less direct racial assumptions have also risen to the surface and have met little to no criticism. When one African American woman reported that her husband, sons, and cousins had the "smarter" diploid combo, she was asked— despite having a clear, close-cropped profile picture of her face next to her entries, showing that she was black—whether her family was Indian or Pakistani (she merely courteously replied that her husband was Ethiopian and then engaged in more discussion of genotypes). When another test taker shared a list of thirty-three "European SNPs for intelligence," no one questioned his haphazard collection of alleles studied in research on memory, brain function, and basic physiological processes, research that happened to be conducted only on people of European descent.[16]

The problem with this kind of crowdsourcing is that test takers share a ton of material that has not been validated, such as unreplicated one-off studies on a certain behavioral phenotype they have read about in the *Wall Street Journal* or the *New York Times*. This is especially disastrous going into the realm of social genomics, since replication studies are for the most part far off in the distance for the majority of the field. Take the test taker who shared a study that claimed "selective pressure on cognition genes due to climate." This person quoted the uncorroborated finding that "COMT Met and rs236330 C are associated with general intelligence with an increase in frequency towards northern latitudes. rs236330 TT is a common genotype among Africans while it is almost non-existent among Asians and Europeans, they are either CC or CT." This test taker went on to paraphrase that "sub-Saharans Africans have the lowest genotypic IQ, that African Americans have higher than them because of better nutrition plus mix with Europeans and that Mongoloid have the highest genotypic IQ."[17] Sadly, this exchange only prompted the sharing of more misinformation on Ashkenazi Jewish intellectual superiority and a debate over whether it would be better to be Ashkenazi or just non-Ashkenazi European.

Through big social media platforms like 23andMe, people are also sharing software and brainstorming ways to get 23andMe to help them merge data on IQ in troubling ways. One test taker who brought news of BGI's IQ genomics data-release gave instructions for syncing 23andMe and BGI data and then suggested "merging of genotypes obtained from different services, and to upload these larger raw files to 23andme as well as to GEDMATCH, HIRS etc." To that, another test taker responded with the good news that "we can migrate our data to the [Personal Genomes Project] and George Church—just emailed him and got his enthusiastic response." Consumers are thus fomenting all sorts of unsupervised uses of genomic software and study data that have heavy eugenic valences and that can be put toward "positive" ("better breeding") eugenic ends.[18]

Born to Rage

Another kind of test out on the market is the aggression test, or "warrior gene" test. Family Tree DNA, the largest personal genomics company in the world, and one that provides genotyping to TV shows and films that feature genetic testing, offers a test for ninety-nine dollars.[19] Under a picture of a Roman soldier that looks like it was taken from the movie *300*, Family Tree DNA asks, "Are You a Warrior? In sports or business, how do you respond to stress? Is the answer in your genes?" The company seems to think so and especially targets men as potential candidates for "reduced function." It argues:

> Because men have one copy of the X-chromosome, a variant that reduces the function of this gene has more of an influence on them. Women, having two X-chromosomes, are more likely to have at least one normally functioning gene copy, and variants in women have not been studied as extensively. The test is now available for both men and women.[20]

What they don't say is that the actual studies that have been conducted on men have not been replicated. Or that many studies have included far too few research subjects to make a valid statement about which populations merit the classification. (But validity doesn't seem to bother them given that they offer women tests despite *no* studies to back tests up.)

Warrior Roots is an American company that creatively combines aggression tests with Y-chromosome ancestry testing to determine a person's paternal haplogroup and then reports what kind of ancient warrior culture that person is descended from. For example, one user with known Irish and Scottish ancestry was told he had his origins in the Scythian culture from the Caucasus. Warrior Roots has cornered the market on aggression testing by targeting mixed martial arts communities and Ultimate Fighting Championship communities.[21]

Aggression tests figure centrally in a number of combination physiological-behavioral test kits that by virtue of including the warrior tests are able to claim to be sociogenomic "lifestyle" kits. For example, the Swiss personal genomics company Genetest's "Lifestyle Gene Test Pack" pairs traits like freckling and alcohol flush reaction with the warrior gene,[22] couching statements like "the so-called warrior gene causes its carriers to be more willing to take risks while simultaneously enabling them to better assess their chances of success in critical situations" in biomedical jargon like this: "'Flushing' is a condition which causes the body to be unable to fully digest alcohol. It causes several symptoms: headaches, nausea, light-headedness, occasional skin swelling and itchiness, occasional extreme drowsiness and an increased pulse."[23]

Companies also couch claims in social science terms, such as in the case of Genetest's sister company iGENEA's warrior gene pitch:

> For a recently published study, scientists at the California Institute of Technology pitted 83 young men against one another in a financial simulation. This entailed them having to gamble and multiply their seed money, USD 25, in different games. Each man had to choose between a sure option (100% no profit, no loss) and a risky option (various risks of loss and chances of profit) 140 times. The findings of the study: carriers of the MAOA-L gene variant were more prone to take financial risks, but only if doing so was beneficial to them. This indicates that carriers of the warrior gene variant are in a better position to make decisions, which are beneficial to them.[24]

Aggression tests are problematic because they bolster a range of unsubstantiated claims about aggression, gender, and race, such as the notion that some populations are "born to rage" or that men with certain alleles have a more unchecked aggression. But they will unlikely disperse any time soon due to the data sharing mentioned above. In fact, even beyond crowdsourcing and user innovations, companies like Family Tree DNA and iGENEA are bridging together to create mega databases for clients to troll and network. For example, iGENEA markets its tests promising, "You also receive permanent and unlimited access to the largest DNA genealogy database in the world, which enables you to find people who share common ancestors with you."[25] Family Tree DNA itself, which boasts its database of "over 700,000 people from over 100 countries," now links to 23andMe and Ancestry.com with its Autosomal Transfers program.[26] These companies form a network of data that reinforces aggression as a stable, innate, gendered, and racial trait.

A Bit More Fit

Fitness and sports make up one of the biggest sectors of the sociogenomics industry and are one way that many biotech companies make quick money on the side. Pathway Genomics, a US-based company known for its cancer pharmacogenomic direct-to-consumer products, for example, which recently launched "Pathway Fit," revoked its early adopter discount within weeks due to its success on the market.[27]

Fitness tests are so appealing to consumers because most have a weight component. Pathway Genomics CCO Ardy Arianpour said, "Pathway Fit was developed with the global obesity issue in mind as a way to help individuals combat the disease. . . . We believe in giving patients the tools to optimize their mental and physical performance in order to achieve a healthier lifestyle."[28] Another US-based company, Phenom Bioscience, offers the "myG:pro nutrigenomics + athlenomics" test kit, which identifies innate snacking tendencies, emotional eating, and uncontrolled eating, among other traits, along with its proprietary CHANGE social media IT platform.[29] These companies play on people's fears about fat, including the worry that no diet will work for them. In a social eugenics fashion, they urge consumers to think of their personal weight in societal terms, as part of a global epidemic and as relational to their friends.

They also use the lure of medical urgency to impress the utility of their tests. For example, Pathway's fitness kit is marketed alongside a suite of more simplified weight and cardio tests. Traits like cholesterol metabolism and pharmacokinetic drug response appear next to feeling full after eating and having a sweet tooth, making behavioral traits look as intrinsic to one's body as life-threatening allergies.[30] The Indian company Xcode's 100 and Life GeneFit test glucose and oxidized fatty acids alongside flexibility and wellness.[31]

Because fitness test makers are dealing with adults who already know their habits and shortcomings, they are tasked with inspiring new concerns in consumers while still encouraging them to optimize their health with better choices. Pathway Genomics' Healthy Woman DNA Insight test exemplifies the delicate tightrope best in its incorporation of further medical expertise.[32] This test, which Pathway specifically aims at women desiring postpartum weight loss, states:

> Healthy Woman DNA Insight tests a variety of genes that influence response to diet, metabolism, and exercise, as well as the propensity to develop certain health conditions and likely response to specific medications. With this information, physicians have the ability to help patients make better decisions to help improve or maintain their overall health.[33]

Another American company, Simplified Genetics, simply promises clarity and self-directed empowerment. Under a giant banner exclaiming "ATTAINABILITY," Simplified Genetics says:

> We don't look for disease pathways or drug interactions, and we are not here to give you bad news. SG translates your genetic instructions to give you your map and the most powerful tools to improve your life; from weight management to picking the safest activities for your child.[34]

Still, nowhere is determinism-meets-optimization so palpable than in the array of athletic ability tests. Sports test makers promise to set clients on a path to athletic success, or to simply help them to exercise right for their bodies. This necessarily implies drawing up laundry lists of physiological mandates. As one Estonian company, Sports Gene, puts it, what these tests determine is the "individual inherent predisposition to succeed" in certain kinds of sports and fitness training based on

> the regulation of blood supply, work capacity and metabolic processes in your muscles; the type of muscle fibres—fast-twitch or slow-twitch; the availability of energy in cells; the availability of constant energy supply in your muscles during exercise; the presence and extent of protection of your skeletal muscles against fatigue; the rate of your muscle growth and availability of energy supply in skeletal muscles; the regulation of the regeneration of myocardial tissue; the consumption of glucose and oxygen in heart and skeletal muscles during exercise.[35]

Or as the US-based multinational company CyGene lists, "muscle growth, cardiac output and . . . bone strength and neurological predispositions" all matter to one's success in training.[36] Yet far from sounding the knell of pessimism, CyGene exclaims: "CyGene's Optimum Athletic Performance DNA Analysis can help you assess what type of sport or event you are genetically wired for and what sports put you at increased risk of physical or neurological injury. Who knows? Olympic success might be in your future!"[37]

And sports tests are just as often aimed at kids, with many companies presenting visual design themes that can reel in adults but with print hinging on tests' life-course predictive power. Companies with names like "Athletic Blueprint" and "Atlas Sports Genetics" dazzle parents with images of sleek double helices and adult athletes winning, while purporting to give "parents and coaches early information on their child's genetic predisposition for success in team or individual speed/power or endurance sports." They emphasize the longevity of results, declaring tests to be useful "later in development with other athletic performance."[38]

Sports test makers also attract parents with promises of safety. For example, Atlas ensures parents that

> genetic testing has potential benefits whether the results are positive or negative for a gene mutation. Test results can provide a sense of relief from uncertainty and help people make informed decisions about managing their health care. For example, a negative result can eliminate the need for unnecessary checkups and screening tests in some cases. A positive result can direct a person toward available prevention, monitoring, and treatment options. Some test results can also help people make decisions about having children. Newborn screening can identify genetic disorders early in life so treatment can be started as early as possible.[39]

This "medicalese" makes it seem like the only risks involved in behavioral testing on kids are those involved with any genetic testing. Companies also connote safety with the ease of sending in tests. Atlas guarantees that it "applies a simple, safe and non-invasive sampling method" and then immediately goes on to say that it's "safe to use on the youngest of athletes," thereby minimizing the significance of risks outside of the test-taking procedure.[40]

Sports test makers also swear by their products by touting their use by athletes and teams around the world. Zybek Sports, a sister company of Atlas Sports Genetics, boasts relations with the NFL, NSCA, Olympic Training Center, and a number of universities and high schools.[41] The Australian company DNAeX claims that its Fit-Test kit has been "used by elite teams and Olympic athletes." Sports test websites provide testimonials and case studies from all kinds of athletes, cricketers, wrestlers, Ironman champions, you name it.[42]

What companies don't make plainly clear is that their work with professional athletes and teams is of a different nature than gene testing. Companies offer run-of-the-mill endurance testing, psychometric testing, and nutritional testing to establish "neuroperformance," "mental resilience and well-being," "athletic decision making," and "cognitive skill." DNAeX, for one, offers an auxiliary suite of psychological aptitude tests with Cognisess Sport, an international brain-training company. Testimonials often merely reflect relationships already built with athletes prior to a company's genetic test making.[43]

A more sinister aspect of sports tests is their framing of race. CyGene offers test results with comparisons to alleged racial norms. For example, in its sample report it situates ACE and BR2 genotype results in a Caucasian-Hispanic-African-Asian matrix.[44] Other companies use stock images of whites and blacks doing different sports to convey racial aptitudes. Genetic Performance, an Irish company, for instance, exclaims, "Now thanks to the latest breakthroughs in genetic testing you can find out what type of exercises and sports you are genetically built for." Their site pictures a white woman competing in a swimming race and a black male NBA player slam-dunking. Its motto "Train smarter" suggests that you'll be better off knowing your racial capacity so you can pursue the right sport for your genes.[45]

Fitness and sports test makers do not seem to understand these implications, as evidenced by the resounding argument you find claiming that tests are a basic right that everyone should have. Genetic Performance, for one, stakes its mission on access, stating, "Our purpose, at Genetic Performance, is to make this ground breaking DNA testing available to all."[46] AIBioTech, maker of the Sports X Factor test, has similarly said, "AIBioTech believes that everyone has a right to their genetic information without a prescription."[47] When the US Food and Drug Administration (FDA) came down on personal genomics companies with multinational presences that sell tests in the United States, something I will discuss in detail in the following pages, companies like AIBiotech and 23andMe fought back with the message, DNA is an inalienable right.

Finding Love

Before getting into the ways that governments have responded to the sociogenomic gene testing industry, I want to present another corner of the industry: relationships. Though there are far fewer tests that center on predicting someone's capacity to secure and sustain love, these tests have been featured the most in national and international news and magazines.

The big sell for many of these companies selling relationships tests is that they offer a social media platform for seeking new partners, or they link up with a preexisting platform that does. The American company SingldOut makes clients sign up through LinkedIn, thus transforming their social media world into a veritable dating pool.[48] The Swiss firm Gene Partner links to dating sites all over the United States and Europe. Thus, these companies can boast access to a deep pool of potential matches just by piggybacking on the ground already laid by other businesses.

Some companies, like fitness and sports test makers, offer a panoply of psychological tests in addition to genotyping services. SingldOut, for one, credits its product with a special "SO Factor" taken from a combination of "genetic and personality factors" that shed light on compatibility.[49] But as with aggression tests, most companies don't find the need to stake the accuracy of tests on psychometrics. The sales pitch of LoveGene, a British company, illustrates this best. It says:

> Have you been using Internet dating to find your other half? Looking for that
> perfect chemistry with someone you'd love? If it never feels entirely right, let us
> help you bring the chemistry back into your dating. Send us a sample of saliva,
> let us do the genetic testing and soon we'll match you up with the perfect date![50]

What LoveGene is implying is that it's time once and for all to kick social dating to the curb. Genes, not the internet, will lead the way to true compatibility.

Chemistry is the watchword of relationships test makers. It's a handy way to biologize the process of finding love. As seen in the quote above, companies like LoveGene imply that unknown biology is at fault for bad experiences out in the field. "The perfect genetic match" is the only match.

No company exemplifies this more than Instant Chemistry, the Canadian dating genetics service that targets both single people and people already in relationships. Instant Chemistry promises that with a few simple genetic tests it can determine psychological compatibility (does your core character mesh with your partner's?), neurocompatibility (are there differences in your neurotransmitters?), and biocompatibility (do your genes suggest serious physical attraction?) all in the interest of predicting whether a given relationship will last.[51]

Of all the tests, these relationships tests portend the most direct form of eugenics: better breeding. Instant Chemistry cuts to the chase, asking, "So you want to know how truly compatible you and your partner are?" It answers that a couples genes will determine if you will stay together or break up, in other words if it's worth it to be married, have kids, and grow old together.[52] In an interview with *Good Morning America*, Gene Partner, whose motto is "Love is no coincidence," called relationships genetics the only true science of partnership. *Good Morning America* celebrated the test for increasing the viability of offspring, stating, "Our offspring will prosper."[53] Companies with dating services boast "awesome members" (as SingldOut puts it, members who are highly educated and professional). Gene Partner takes a more biological stand. It argues from an entirely medical and eugenic standpoint that "the probability for successful and long-lasting romantic relationships is greatest in couples with high genetic compatibility. . . . With genetically highly compatible people we feel that rare sensation of perfect chemistry. This is the body's receptive and welcoming response when immune systems harmonize and fit well together."[54] Companies like Gene Partner promise that relationships based on genetics will last because sex will be more enjoyable, thereby boosting fertility. Genetic testing is thus a no-brainer. It's the only path to mating success and the establishment of viable future generations.

The Wild West of Sociogenomic Tests

You might be wondering what the situation is with regulating these companies. The answer is complicated. While there have been attempts to regulate the personal genomics market in many countries, overall there has been very little control of company activity, and no oversight of sociogenomic tests at all.

In the United States, the FDA has debated the selling of direct-to-consumer genomics for several years; however, the fight has been largely over medical health information and not the behavioral traits described above. The FDA's most recent major injunction against the personal genomics industry occurred in 2010 when the agency required seventeen companies to stop selling home medical kits. Graceful Earth's Alzheimer's test, SeqWright DNA's Genomic profiling tests, Interleukin Genetics' Inherent Health test, DNATraits' Ashkenazi Jews Disease Panel, CyGene's Metabolic Health Assessment, Consumer Genetics' Asthma DNA test, Matrix Genomics' Breast Cancer test, Genetic Testing Laboratories' DNA predisposition test, Sequenom's SEQureDx test, Entero-Lab Ref Lab's gene celiac disease test, BioMarker Pharmaceuticals' Gene Essence test, DNA Dimensions' predisposition DNA test, HealthCheckUS's celiac disease test, and easyDNA's genetic predisposition test all were given cease-and-desist notifications.[55]

In 2013 the FDA, targeting 23andMe specifically, barred any delivery of health data from genome scans. One year later, the agency began drafting new guidelines for laboratory-developed tests and direct-to-consumer genetic tests for all companies.[56]

In February 2015, the FDA revoked its medical ban for one specific carrier test on an extremely rare syndrome in children called Bloom syndrome. This approval opened discussion for further authorizations for specific tests, and the expansion of device and testing proposals from companies like 23andMe.[57] After that, by "working within the system," as 23andMe CEO Ann Wojcicki states, the company resumed providing health data to users. The health data approved by the FDA correspond to very specific (and many rare) disease screenings for whether a test taker's child will have the disease. 23andMe hopes to "reignite" its membership with this update, claiming it "redesigned the whole experience."[58] In 2015 alone, the company surpassed one million users and raised $115 million from investors, a very successful rebound after being "out of the game" for two years.[59]

It's important to note that also in 2015 the FDA launched "precisionFDA," its own commitment to Obama's "precision medicine initiative" drive to create a collaborative precision medicine community.[60] The world's first open-source platform, precisionFDA is a cloud-based databasing system for genomic companies to share data and methodology.[61] The purpose of precisionFDA is to create a government-sponsored share economy in which next-generation sequencing and diagnostics can be community vetted at their inception, and for the FDA to be able to "cross validate their tests or results against crowd-sourced reference material."[62] Interestingly, despite the FDA's goal of playing a more active role in oversight, 23andMe and the medical company Counsyl have been the two personal genomics companies most involved with the platform.

The FDA has walked a delicate tightrope of increasing oversight and allowing the market to develop on its own, because the agency to date lacks legal authority over laboratory-developed tests. In 2015, the FDA proposed a draft of potential oversight, but it has not yet made any actionable formal rules under the US Administrative Procedures Act.[63]

There are also loopholes in its current oversight program. As of now, the FDA permits Investigational Device Exemptions for clinical research before a study is initiated, for devices not yet cleared by the government. However,

Investigational Device Exemption applications are not standardized. They often are simply required to report issues of risk, labeling, supplied consumer information, and institutional review board details.[64] The FDA maintains that it is open to the development of new applications, promising that "unlike the enforcement discretion practiced in the realm of commercial genetic tests, the agency has never discriminated between [laboratory-developed tests] and kits in exercising this authority on the use of genetic devices within clinical research."[65] Yet the agency still remains hesitant with approvals for hereditary breast cancer and pharmacogenomic purposes.

The only companies selling sociogenomic tests that have received FDA warnings are CyGene, 23andMe, and Pathway Genomics. However, these letters have been about their medical health tests. Looking closer at one of these cases proves telling. In September 2015, the FDA issued Pathway a letter for its CancerIntercept Detect test, which it was marketing without governmental clearance. The FDA warned Pathway that it was unable to find any approval or even listing for Pathway's direct-to-consumer screening devices. The FDA also could not find any evidence that the test had been clinically proven for efficacy. In its review of the company's citations, it found that the company had not substantiated its "expansive claims" of screening capacity, suggesting that companies print false claims about their products.[66] Despite finding such grave fault with Pathway's cancer product, the FDA did not investigate its lifestyle devices.

In the past couple of years, the FDA has monitored several companies closely. One company is its partner in precisionFDA, Counsyl Genetics. Though this company provides several controversial cancer screening tests, the FDA has approved a number of its tests for specific diseases, such as Bloom syndrome and Krabbe disease, and for noninvasive prenatal testing.[67] The FDA has also monitored GenebyGene, a company involved in a lawsuit with Myriad over its breast cancer gene test. Despite US regulations and a lack of FDA approval, GenebyGene has moved to an international market.

The FDA has also kept an eye on companies that are expanding into the sociogenomic realm. InVitae, a company that has offered breast cancer gene sequencing for many years, has been developing "education programs" to increase its clients' use of family genetic testing. Interleukin Genetics, a company targeted in 2010 for patent infringement by Australian company Genetics Technology, was granted a European patent for a weight management genetic test in April 2015, thus acquiring major markets in Eurasia, Japan, Mexico, Russia, and New Zealand.[68] Stanford Sports Genetics, a sports-related genetics company, offers gene screening for "athletic prowess" in collaboration with 23andMe.[69] And then there is Helix, a company launched in 2016, which markets its therapies as "digitized genomics."[70] But many companies feel unthreatened by the FDA's attention. As Genelex, a pharmacogenomics company that markets over twenty pharmacogenomic tests and that was the founder of Family Tree DNA, has said, the FDA is simply too slow to keep up with the market.[71]

The United Kingdom appears 180 degrees opposite of the United States. As of now, it has no formal regulation, though there are bioethics nongovernmental organization agencies, like the Genetic Testing Network[72] and UK Human Genetics Commission,[73] which monitor personal genomics activity. In 2009, the UK's Science and Technology Committee of the House of Lords published a report on genomic medicine calling for "a voluntary code of practice, which encourages providers to be open about tests' limitations and enables consumers to make informed decisions,"[74] and a Department of Health website for consumers containing up-to-date and comprehensive information about direct-to-consumer genetic testing companies and the tests they offer.[75] In 2010, just before being disbanded by Parliament in a sweep of elimination of quangos (commissioned governance bodies that are formally outside the government), the UK Human Genetics Commission drew up the Common Framework of Principles, a set of guidelines put in play to protect consumer interests and to "promote high standards and consistency" between companies.[76] All genetic tests in the United Kingdom fall under the EU's 1998 In Vitro Diagnostic Devices Directive, and the UK's Human Tissue Act of 2004, which safeguards people from DNA theft—"the collection or analysis of individuals' DNA without their consent."[77] However, these laws do not provide specific oversight of personal genomics direct-to-consumer tests. Therefore, in 2013, when the FDA banned medical direct-to-consumer tests, 23andMe expanded to the United Kingdom seeking new markets. The UK's parallel regulatory agency, the Medicines and Healthcare Products Regulatory Agency, received 23andMe with open arms, simply advising consumers to use tests with caution.[78]

In fact, in most European countries, personal genomics is not regulated. Switzerland, France, Germany, the Netherlands, and Portugal all have some kind of policy about providing genetic counseling or medical supervision, but as bioethicists across Europe have shown, government regulators only take issue with the quality of information given to clients. As one team of bioethicists has argued: "In these countries, the underlying premise is that individuals should be given the opportunity to make their decisions freely and this should be based on adequate information about the limitations of [direct-to-consumer] genetic tests and their (physical, psychological and social) implications."[79] The Netherlands is a case in point. It has no specific legislation on direct-to-consumer tests, although the Dutch Act on Population Screening aims to protect individuals from programs that are deemed a threat to health, such as direct-to-consumer tests aimed at cancer and incurable diseases.[80] The act was not intended to regulate the personal genomics industry, and as such it has not limited the influx of companies like 23andMe and the many others presented above. While Dutch scientists voice skepticism of personalized genetic testing, little has been done to oversee the industry.[81] Germany too passed legislation in 2009 banning direct-to-consumer genetic testing without the presence of a licensed medical doctor and a patient's informed consent. However, like all regulations elsewhere, the law does not ban the use of tests purchased abroad.

China is the only country that is putting strict sanctions on the use of genetic tests. In February 2014, the government put a moratorium on tests used in hospitals and clinics in an effort to crack down on prenatal genetic testing. Though the policy does not specifically address direct-to-consumer testing, it has curbed the development of personal genomics in China's pharmacogenomics giants and its big test makers, BGI and Berry Genomics.[82] Since the ban, the Chinese government has approved Berry's prenatal test made with partner Illumina.[83]

Other than China, most countries around the world have not produced injunctions against the personal genomics industry and its sociogenomics subsidiaries. Many have produced policies counter to those more stringent policies in the United States and Europe. For example, Australia, a state that has been open to direct-to-consumer biotech markets, has given wide berth to its major genetic test makers. And in September 2014, Australian courts upheld patents on BRCA1 that the US Supreme Court shot down in its 2014 verdict.[84] More often than not, countries outside the United States and Europe have provided a haven for American or European companies interested in expanding into regulated markets. BGI, for example, opened a next-generation sequencing facility in Singapore in 2012.[85] Similarly, Germany's Merck expanded operations to Israel in June 2015.[86]

Several international regulating bodies could in the future make a bigger impact on the personal genomics industry. The Organisation for Economic Cooperation and Development (OECD) set guidelines for its eighteen member countries.[87] One of the report's key conclusions was that labs in all countries should share the work of data analysis, sending samples back and forth across borders. The OECD's major recommendation was to establish accreditation standards to ensure quality across international networks. The European Union has provided the following: Directive 95/46/EC on the protection of individuals with regard to the processing of personal data and on the free movement of such data; Directive 2000/31/EC on certain legal aspects of information-society services, in particular electronic commerce in the internal market; 93/42/EEC with 2007/47/EC, the Medical Devices Directives; Directive 97/7/EC on the protection of consumers with respect to distance contracts; Directive 2006/114/EC concerning misleading and comparative advertising; Directive 2005/29/EC concerning unfair business-to-consumer commercial practices and Competition Law.[88] Again, none of these directly address personal genomics direct-to-consumer testing.

Tests That Travel: More on the Global Nature of Sociogenomics

Market research on the direct-to-consumer genetic testing industry agrees that what we are dealing with is a multibillion-dollar global industry.[89] Tests are in demand due to the rise in chronic illness and the aging demographics around the world, and also the widespread availability of information on tests. As analysts report, genetic testing is the "most rapidly expanding segment of the molecular diagnostics market worldwide," one that is transforming from a service-based market to a "product-driven" market.[90]

The United States is the largest market for tests, projected to reach $2.2 billion alone by 2017. America has been an especially fertile market because the medical community has come out in support of genetic testing and public health has integrated genetics into its operations, especially in the area of maternal and newborn screening around rare diseases like sickle cell anemia and cystic fibrosis. Testing around cancer, including predictive testing on alleles like the BRCA 1 and 2 genes, have also made genetic testing commonplace in healthcare and medicine. And the convergence of ancestry testing and medical testing has expanded the scope of the market.[91] In fact, lifestyle improvement has been reported as one of the most common reasons that people buy tests. One survey of over one thousand consumers found that 94 percent of respondents purchased tests out of curiosity, with most interested in learning about potential future health and disease.[92]

After the 2013 FDA ban on medical tests, there was less reporting on actual sales. But personal genomics companies, which still sold ancestry testing and "raw" genetic scans, nevertheless projected growth.[93] For example, 23andMe reported that more than 500,000 Americans used the kits before the FDA ban, but that it expected to get to 800,000 customers in 2014. And as mentioned, reports in 2015 show that 23andMe had indeed reached one million customers.[94] Similarly, despite dips in consumption, Genomic Health, a marketer of cancer tests, reached over 500,000 customers in 2015. Now the company is expanding overseas to pursue a "$2 billion market opportunity outside of the US."[95]

Indeed, in 2014, 23andMe introduced 108 health-related reports and ancestry testing to the Canadian market with no pushback. Canada permits the sale of direct-to-consumer health information and has no regulations for specific traits. Even before the 2014 expansion, an estimated 20,000 people had bought tests.[96]

And in the United Kingdom, where 23andMe is focusing much of its efforts using a combination of tests for dozens of health conditions, inherited conditions, and traits, and where the National Health Service has been sequencing as many British nationals as will participate in their 100,000 genomes project, new markets abound for the range of companies offering DNA tests.[97]

23andMe has also continued to market all its tests directly to consumers in Denmark, Finland, Ireland, Sweden, and the Netherlands, and to another fifty-six countries that only permit nonmedical information. In these kits, the sociogenomic tests and ancestry tests are face forward.[98]

Finally, markets in the Middle East and Asia have opened up to the global exchange. In the United Arab Emirates, for example, there have been great efforts to secure Dubai's position as a biotech hub, with its establishment of the DuBiotech Park and support of Eastern Biotech, a homegrown direct-to-consumer contender.[99] In China, where bans were conducted in 2014, there has been no significant decrease in sales from prior years, and the government has committed to becoming a leader in gene editing.[100]

All this shows that the reason why tests are so successful is because they are born into a world where international partnership and global horizons reign. There is no binding international regulation on personal genomics companies. And even when there are national stipulations, companies can just hop a border or go transnational if they want to expand their protocols and markets. Companies indeed thrive off of reinventing themselves as international. Pathway Genomics, for one, boasts itself as the "Global Clinical Genetic Testing

Lab" and provides services in more the forty countries, and when regulated in the United States companies like 23andMe have expanded services internationally.[101] And company leaders often have their hand in a variety of markets around the world. This is perhaps in part due to the fact that the majority of personal genomics companies in existence have board members, founders, and chief officers who serve similar positions in different companies. Patrick Chung, a board member of 23andMe, for example, is an investor for Crowd-Med and previously sat on the board of directors for Rock Health and Euclid Analytics.[102] Esther Dyson, also on the board at 23andMe, serves on the advisory board for Medivo and Voxivo, both health-related enterprises.[103] The CEO and founder of Pathway Genomics, Jim Plante, also founded SmartDrive and was recently appointed to the Industry Trade Advisory Committee.[104] At Interleukin Genetics, board member Dayton Misfeldt additionally sits on the boards of Sunesis Pharmaceuticals, Presidio Pharmaceuticals, and Next Wave Pharmaceuticals.[105]

This all begs the question of whether, as Nikolas Rose puts it, biology in the postgenomic age is opportunity (as opposed to destiny).[106] Does the light hand in regulation paired with a booming global market in sociogenomics mean that people are getting the chance to do more with their DNA? Does it mean that they can optimize freely? After all, governments do not currently regulate the use of personal genomics tests; they merely regulate their provision. Any individual can go online or to another country to buy tests. People can bring home as many tests on as many traits as they want and gift them to as many people as they like. Tests aren't going anywhere. They're going everywhere.

To all these questions, I reply with a firm no. While people may have the opportunity to buy access to computer readouts about their personal DNA sequence, they are not given that information without a heavy dose of interpretation. And as we've seen, tests are being pushed along with all sorts of monitions. In the end, sociogenomics in the market is just as deterministic as in the media. Yet its potential to encourage eugenics makes it all the more threatening. It is time that we create meaningful policy and practice before it's too late and the sociogenomic paradigm's nature-first, genes-first, essentialist framework is so deeply rooted that we can't extricate ourselves from it.

CONCLUSION

Where are we with all this sociogenomics? As asked in the beginning of this book, is a new and unique field of research forming? If so, what does its disciplinary terrain and expertise look like? Secondly, is the presence of this new science promoting geneticization? And what are the implications of the science for societal notions of human difference, such as race, gender, and sexuality?

This analysis has shown that the answer to the first question is *yes*. While there isn't a formal program in sociogenomic science, no academic department or doctoral degree to that end, there is a growing movement in the sciences toward a combined approach to studying gene-environment interactions, and that approach is no longer the domain of any one natural or social science. Instead, it is congealing in the form of a fast-institutionalizing community of scholars, an autonomous field, that is increasingly visible in its own right.

As such, sociogenomics isn't a mere scientific intellectual movement, fighting its way to a place of legitimacy within the sciences. It doesn't face significant criticism from other sciences. Rather, it is enjoying a great reception from mainstream genomics and the wider intellectual community. The media regularly reports on sociogenomic findings, and it does so in ways that guarantee the paradigm's legitimacy. Moreover, this science has moved into the private sector and the limelight. Its approach to the very essence of life is increasingly a normal part of our world.

The contours of sociogenomics, and its leading proponent field of social genomics, prove to be entirely novel and multidisciplinary on all registers. It brings original forms of cross-pollination within the natural sciences in the form of new research agendas that bridge fields like epidemiology, evolutionary biology, genetics, and health science. It also brings an even greater deal of novel cross-pollination within the social sciences between disciplines that typically are to a great deal siloed off from one another, such as sociology, economics, and political science. But of course the most original aspect of this form of science is its wedding of natural and social science disciplinary approaches and interests. It is transdisciplinary in ways that previous sciences, even sociobiological and gene-environment ones, have not been. At stake for it isn't a specific domain of expertise, but rather a capacious interest in being the mouthpiece for anything social and genetic.

In some sense sociogenomics is fulfilling its promise of revolutionizing science as we know it. But in other ways it's continuing the practice of what I'd call "imbalanced interdisciplinarity." Is it possible that a science so *multi* can be so *mono*? Yes. And this imbalance, which has things slated in favor of genetic supremacy, has important implications for how the science impacts the public.

Imbalanced interdisciplinarity in favor of the already dominant genetic side of things means that social behaviors are indeed being geneticized as this multidisciplinary science engages them. To the second question I've asked, then, I give another wholehearted, affirmative *yes*. Yes, sociogenomic science is eating up social phenomena and spitting them back out as biological processes. Yes, it is casting previously nongenetic traits and behaviors in genetic terms.

In fact, all of my analyses here point in one direction. Whether looking at a tight and narrow ring of scientists and their foundational methods, or institutional networks that are now sponsoring the science, or even organizations and social institutions far afield in the public, we see not only a new emphasis on dealing in genetics but also an encroaching genetic determinism. It's not that individual scientists—the geno- poli-, socio-, and econ folks who are leading the way—buy into genetic determinism. Far from it, they want nothing more than to complicate oversimplistic notions of genetic supremacy.[1] They want to show the sciences and the world that the environment matters to our biology, health, and life outcomes. And that policy matters, justice matters. But because institutional sponsors are riding the high of new interdisciplinary approaches, and thus ignore the specific ways in which the environment factors into research so thinly, the race for genetic innovation is the focus. And because scientists consequentially can run with gene-gene analyses and research questions that ignore the substantive social context, it is all too easy for the media to get a hold of "gene found" studies and simply report the genetic findings, or for companies to capitalize on lay beliefs in the essence of DNA. From there, it's only a skip and a jump till experts in the public are engaging the new science in ways that promote that very narrow-minded genes-first form of interpretation.

In other words, the notoriety for molecular causes creates a kind of "legitimacy feedback loop" for the field and for its sponsors, casting its members as the frontiersmen of science who are tasked with finding more genetic culprits for the social behaviors we care most about, and its supporters as the risk takers who have made transdisciplinary science a reality. Though I didn't find that others further afield are yet gaining accolades for their participation in or application of sociogenomic science, I wouldn't be surprised if the prestige extended outward to others too in the future, such as experts in the broader public who use tests, or the organizations and entities who become institution-wide early adopters. We might find ourselves living in a world where institutions are incentivized to use social behavioral genetic screens, either through monetary benefits or nonfiscal rewards. Individual citizens may also be lauded for participation in gene-based training or prophylactic anticrime programs one day.

In terms of how we may want to conceptualize or theorize this, it's critical to capture the values behind the process. With sociogenomics, geneticization isn't just about medicalization by genetics or about scientists gaining cred by allying with a publicly prominent field. It's about the rise of a wider social paradigm for managing health and the body, populations and the body politic, that plays on the ethical aspirations of experts and laypeople alike. Geneticization is, to use Nikolas Rose's term, a form of ethopolitics. It is deeply connected to our most cherished values. Some of these values reflect long-standing liberal and humanistic notions of individual equality and freedom, values that may have a stronger tradition in the places where social genomics has arisen but that circulate through the global marketplace, infusing a variety of cultures and polities. Others of these values reflect more recent ideologies of personal responsibility and health welfare, such as ideologies around the need to manage risks and the necessity for tools to do so. Scholars of geneticization must mind these ethopolitical dimensions as they elaborate its processes and mechanisms.

And with molecularization, especially the advancement of genetic models within the halls of science, sociogenomics shows that it is driven not just by the utility of genetic models in the study of specific research problems, but also by larger structural shifts toward interdisciplinary knowledge production. It is this broader social context of incentives that makes genetics not only appealing but also imperative to scientists coming up in the new millennium. And in fact, as disinterested as these larger structural shifts may seem, they too are tied to the value systems at play. They are part and parcel of today's ethopolitics, which in this postgenomic age equate an ecumenical approach to research with better knowing.

These developments in geneticization and molecularization suggest that scholarly accounts of organizations, professions, and fields in the postgenomic age will be inadequate without attention to the ways that ethopolitics imbricate the reward structures and institutional dynamics that sustain them. We tend to think of organizations, professions, and fields as permeated by turf wars where collectives and individuals compete for scarce resources, or vie for different kinds of capital. This may be true, it but isn't as value free as we often state. While experts in the study of organizations, professions, and fields have acknowledged the passion within the struggle, I believe we can go further.

The scientists, policymakers, funders, and experts in the broader public depicted here aren't just passionate insofar as they are fighting a boundary war. They are passionate about the ethical implications of their work, and in ways that link to the core of who they see themselves as and the world they want to create. The simultaneously heterodox and orthodox ways that sociogenomic researchers achieve their goals, like targeting top journals with great influence and infiltrating leading genetic research institutes while attempting to remain under the radar in their own departments, are enacted in the efforts toward something bigger. The same goes for the boundary crossing in which experts beyond the science engage. At the current moment, in this postgenomic age, genetics is seen as a lifeline to a better world by so many entities and stakeholders, and activism is viewed as every person's responsibility—everyone is at the frontier of our future and must act. Thus, geneticization and molecularization are not as risky or groundbreaking for organizations, professions, and fields as it may seem. Only an analysis of such entities that explores ethopolitics such as that belonging to the ethos of the postgenomic age will ascertain these fundamental social processes.

As for the second question, the novel social processes that I describe here matter a great deal to how we as a society conceive of human difference, dissimilarity, deviation, and disparity. In the science described here, differences like race, gender, and sexuality are assumed to be essential biology. Race is treated like something to be controlled for, something to be managed with genetic technology. Gender differences are almost wholly confounded with sex differences, ignoring the social processes by which gender identities are formed and the biological or health impacts those identities have on people. Sexuality is problematized to a much greater degree, but only when it comes to sexual orientation. Therefore, these highly deterministic ways of dealing with human difference make for a world in which human difference is defined by genetic characteristics.

As a result, the greater science community and public are left to believe that social disparities and inequities are in the blood, "under the skin" as one genosociologist put it, but only insomuch as we're talking about inborn biology. It becomes all too easy to view the persistent social injustices of our society, as well as health inequities, to use bioethicist Dorothy Roberts's insightful terminology,[2] as genetically induced. As I have written elsewhere many times, many governments around the world are attempting to eradicate health disparities by dumping millions, if not billions, into health research. But increasingly this support is being channeled to genetic research. Without a gene-environment approach that turns the tables on genes-first science, we will continue to miss our target for upending injustice and inequality, social realities that hold origins in the structural environment far outside the corporal or cellular body.

Again, conceptually and theoretically speaking, this points to a new kind of construction of human difference, one that is even different in some ways from the genetic science of only a decade ago. In the space of early genomics, difference was grappled with directly and scientists fought to win over the public about their own definitions. Now in the postgenomic age, difference is being investigated and characterized on a more subtle level, even as scientists champion inclusion and social equality. As this book has shown, researchers have high hopes that their science will be a driver of racial, gender, and sexual justice. Yet in uncritically importing genomic methods of stratifying by race and sex, and then linking these traits to things like aggression, promiscuity, and intelligence, essentialism is reinstated.

Furthermore, the genomics of old avoided studying behavior and social outcomes, because its scientists greatly feared a return to eugenics. Now sociogenomic science is facing behavior and social outcomes head-on, and the specter of eugenics isn't deterring it. Social genomicists hope to help people capitalize on their own bioplasticity, by producing research that can convince policymakers and other experts in the broader public to tailor policy and practices to their individual needs. In essence, better breeding is bypassed for better socialization.

But with a nature-first approach that supports a nature-versus-nurture dichotomy, it will be difficult to avoid impressing more genetic determinism on society. Thus, while the strategies of sociogenomic scientists are different from anything we've seen, the outcomes may not fall far from the tree. I write this at such an early stage in the field's development precisely as a wake-up call to its founders, who I believe have the very best intentions for a just future.

This is why it is critical that neither a small group of scientists nor an even smaller group of funders single-handedly lead the way forward. Natural and social scientists need to take heed of these developments and respond. Laypeople must also get involved, not merely as research subjects or consumers but as stakeholders in the building of this science. Together we must pressure funders and policymakers to require experts in nonsociogenomic science, ethics, and policy as well as members of the public to be at the table as funding priorities are determined and research agendas are cast.

Frontdoor to Eugenics

How the truth about human difference figures into genetic science, and science writ large, is today important in a way that could not have been foreseen even a year ago. Over the past several decades, historians of science and social scientists of race and medicine have kept alive a number of debates over the potential for new genetic sciences to bring about a resurgence of scientific racism, and even possibly a full-fledged eugenics. To many of their outcries scientists of the genome have responded, "That was then. This is now." Even sociologists of science like myself, who witnessed the tsunami of antiracist values hit the shores of science with lightning speed, began to feel optimistic that scientists might steer us away from the chance for discriminatory eugenics programs based on the circulation of new technologies.

In 1990, Troy Duster wrote *Backdoor to Eugenics*,[3] a comparative ethnography of different negative and positive eugenics programs of the twentieth century. He warned that eugenics was still alive, entwining with everyday racial classification processes, to form a racial eugenics that promotes social inequality. Some examples he gave were the US military's attempt to genetically screen pilots for sickle cell anemia, and state-versus-community-based screening programs for diseases in the United States and rural Greece—programs that relied on genetic determinants of race and ethnicity to administer screens under varying degrees of coercion and with varying consequences in leaving the public with deterministic notions of race. What he showed was that the firm belief in science and society that race was in our genes, and its growing reflection in new genetic discourses, was priming the world for a new racial eugenics program—a softer, better-intentioned eugenics program, one slipping in cautiously if not gingerly through the backdoor, but a racial eugenics program nonetheless.

Today something big has happened that has flung open all doors and brought eugenics to bear as a central matter for our collective future. A new technology has hit the scene: gene editing of unwanted variation.[4] Scientists can now edit in or out genetic nucleotides that cause disease, promote health, or lead to the social outcomes we desire most. Eugenics has entered through the front door, pulled up a chair, and is plopped down right in the middle of all the new efforts for better, fitter bodies and minds.

This technology comprises various cutting and pasting techniques that rely on chemicals with nifty names like CRISPR and Antisense, zinc fingers and TALENs. These chemicals allow scientists to nip and tuck their way through the genome, and even the epigenome, programming our bodies to do what we as a society deem critical to a good life. As I write, companies around the world are developing therapies that can kill cancer, take out Tay-Sachs, and potentially save the planet from the litany of common chronic diseases that cut short our lives.

Along with the power to edit out disease comes the power to edit for enhancement. In December 2015, I attended an international summit hosted by the US National Academies of Science, Engineering and Medicine, the UK Royal Society, and the Chinese Academy of Science, in which researchers seemed to agree that editing for enhancement purposes should not reach the clinic until further research had been done.[5] Yet researchers also suggested that there would be no way to draw a bright line between medical and enhancement aims going forward. Geneticist George Church, of the famed Personal Genome Project and the BRAIN Initiative, gave the example of Alzheimer's. How can anyone distinguish between healthy brain function and beneficial cognitive performance? What about greater lung capacity or muscular tone and the ability to stave off physical illness?

In many conversations with scientists at the forefront of this technology, I and several other social scientists raised the issue of equity. If intelligence measured by IQ is linked to income, not to mention fair treatment by all sorts of gatekeepers who hold the key to quality-of-life issues like obtaining a home loan or getting into a good school, how could we deny a person's right to a therapy that could bump up their IQ even just a few points? What about height, which has also been shown to lead to better life outcomes? What about the traits we are most uncomfortable about changing, such as skin color and hair texture? With the ensuing racial violence committed by whites against blacks, could we really in all conscience deny a parent pigmentation-altering therapies that might just mean the difference between life and death?

So eugenics comes in through the front door. But backdoor questions still remain: In a world where social behaviors are being cast in terms of race, gender, and sexuality, how will we avoid creating regimes that edit on the basis of them? How will we avoid a eugenics that targets social groups if it is their genes that are marked as better or worse?

Bioethicist Tetsuya Ishii has predicted that despite the fact that some countries will try to regulate gene-editing markets, just as with personal genomics, medical tourism will prevail. As it is, the business in these technologies is shooting through the roof in the countries that don't appear likely to regulate the market—the United States, United Kingdom, China, and Japan, to name a few. In the months preceding the Gene Editing Summit, companies were raising billions to develop gene-editing platforms.[6] Projections had the market doubling by 2019.[7]

The point I made there and elsewhere, and that I will make here again, is that good intentions are never enough. We may have the best intentions for our science, but without proper policy misuses are almost guaranteed. The scientists doing this work, the institutional sponsors supporting it, and the early adopters applying the science all hope to achieve better social and policy environments to help those who need it most. But the outcome of all this will certainly set us back centuries in terms of stereotypes and social relations, and propel us forward into a dystopian future where injustices against social groups are left unaddressed save for the application of genetic technology to physiological traits.

Sociogenomics into the Future

The bigger question that anyone reading this will want answered is whether sociogenomics is taking us to a more ethical future? Are we better off with this paradigm that originally melds the social and the natural to forge innovative pathways to understanding human being? It is probably not surprising that my answer is a qualified *no*. No, we are not heading into more just and ethical terrain simply because of the originality of the science, nor are we moving in a good direction for society simply by virtue of a more complex form of innovation. But I do see potential to open up a dialogue that can make a better synergistic paradigm, one that is not imbalanced but rather is equitable at its core.

I am talking about an even more transdisciplinary social-genomic health science, one that draws on the strengths of large-scale collaborations between genomicists, health researchers, social genomics researchers, and social scientists who are not studying the genome. I'm talking about studies where the environment is measured just as thoroughly as the genes, not in absentia, where surveys and DNA samples have been deposited into a database preanalysis, but in real time as genetic recruitment and analysis unfolds.

I am also envisioning studies in which the public is involved as research stakeholders, directing what kinds of questions we want to ask. All too often findings are reported via the news media and companies run with innovations to market; then everyday people hear of research applications via their own movement through organizations and institutions or via commercials selling a particular product. The public most certainly needs to be involved in debates over what we should do with gene editing, with enough time to create policies for clinical applications.

One big takeaway from all of this, which I stress again, is that we can't leave it to a small group of scientists and funders to direct science priorities alone. The sociogenomic paradigm is all about ethics. It's all about best intentions. Sociogenomics is the humanization of cold, dry laboratory science. It is the personalization and politicization of the knowledge of our deepest life essence. But when that ethics comes out of the experiences of a select few, it is bound to be partial, biased, and imbalanced, even if enacted under the auspices of broadly shared values.

To conclude, sociogenomics, with its capacious way of applying to everything human under the sun, demands critical attention now. Despite involving a blurring of all these institutional and ideological boundaries, sociogenomics is still tilted toward intrinsic *explanans* for increasingly extrinsic *explanada*. So while it may not appear essentialist or deterministic, it's still pushing just that kind of agenda and it is extending this agenda to characteristics that don't necessarily merit the brand of genetic causality.

Indeed, it is the capaciousness of sociogenomics that is most threatening if not addressed immediately by experts and stakeholders in the general public.

As I mentioned at the opening of this book, sociogenomics is in science, it's in governance, it's a part of the expert's toolkit, and it's increasingly present in the public sphere. It bows to no one discipline. It resides in no single sector or market. It is global. It's local. It's about the population. It's about you. We must address it with an equally plural, multiple, complex set of voices so that we can harness the science for true social good.

I hope that the information in this book will encourage those who are leading and supporting the science to consider a more meaningful public engagement, and that experts and non-expert readers will also think about getting involved via the science policy forums in their own communities. Finally, I'd like to encourage all of us to continue to rethink and reshape the ways research subjects participate in studies. By opening channels of communication for subjects to voice their own views and ideas about how to group, interrogate, and describe, we will bring about a more equitable human science.

ACKNOWLEDGMENTS

I first and foremost thank the scientists who participated in my research. Without their openness and honesty, I would never have been able to capture these important developments in science and society in the postgenomic age. I also thank the many other experts who shared their experiences working in the realm of sociogenomics. Their efforts and interpretations are a critical part of the trajectory traced here.

Thanks are due to the sponsors of my research and the various scholarly communities that hosted presentations of it. Among them are the National Science Foundation Science and Technology Studies Program, University of California Center for New Racial Studies, UCSF Clinical and Translational Science Institute, Center for Transdisciplinary ELSI Research in Translational Genomics, American Association for the Advancement of Science, Genetics and Genealogy Working Group, and Critical Race Theory and Health Sciences Working Group. Within these organizations, I am particularly indebted to Nina Jablonski, Terence Keel, Barbara Koenig, Osagie Obasogie, and Howard Winant.

I am deeply indebted to my research and publication teams, my research team manager and lead research assistant, Chris Hanssmann, senior analysts Melanie Jeske and Rosalie Winslow, and Charles Clongier, Megan Dowdell, and Noa Nessim. The skill each contributed, and the collegiality they built into the collective research experience, made conducting this study enjoyable and exciting. Kate Wahl and her brilliant team at Stanford University Press, including Stephanie Adams, Olivia Bartz, Anne Fuzellier, and Marcela Maxfield have made publication a true pleasure. Thanks especially go out to Jeff Wyneken for his expert copyediting.

I would like to thank my colleagues in my department at UCSF. Shari Dworkin, Howard Pinderhughes, and Janet Shim have been true friends to me through the research and writing process. I would also like to thank Ruth Malone, who served as Chair throughout much of the research period. All of these colleagues have gone the extra mile to read drafts, assist with grant proposals, and mentor me on the road to tenure.

My deepest gratitude goes to my colleagues further afield who are not only my interlocutors; they are my inspiration. Ruha Benjamin, Catherine Lee, Ann Morning, Alondra Nelson, Aaron Panofsky, and Sara Shostak are all dear to my heart. My mentors Phil Brown and Troy Duster are also due thanks for their continued support. I couldn't dream of a better community of scholars to grow up in.

Lastly, I would like to thank my friends and family. The New York crew has spread to Los Angeles and San Francisco, but we are closer than ever. I feel blessed to have the best of friends to carry me through life, making every moment rich, meaningful, and full of emotion. I am most grateful to my chosen family, the Woodburys, and my family in Indonesia and the United States. I especially would like to thank my mom and dad, to whom this book is dedicated. Their love has been the root of my love. It is the foundation for what I give to the world. I lost my papa and my dad during the making of this book. I miss them so much.

Most of all, I want to thank my husband, the love of my life, Nick. Every thought, every creation, is connected to him. His kindness and grace are my joy.

NOTES

Introduction

1. Ewen Callaway, "'Gangsta Gene' Identified in US Teens," *New Scientist*, 2009, www.newscientist.com/article/dn17337-gangsta-gene-identified-in-us-teens/.

2. "Single Forever? It's in Your Genes," *Independent OnLine*, 2014, http://sbeta.iol .co.za/lifestyle/love-sex/relationships/single-forever-it-s-in-your-genes-1784218.

3. "Transporter of Delight: Happiness Is in Your DNA; And Different Races May Have Different Propensities for It," *Economist*, 2011, www.economist.com/node/21532247.

4. "Sexual Orientation Not a Choice, but Is Influenced by Genetics." *Free Press Journal*, 2014, www.freepressjournal.in/sexual-orientation-not-a-choice-but-is-influenced -by-genetics.

5. See, for example, C. Rietveld, S. Medland, J. Derringer, J. Yang, T. Esko, N. Martin, H.-J. Westra, et al., "GWAS of 126,559 Individuals Identifies Genetic Variants Associated with Educational Attainment," *Science* 340 (2013): 1467–71; N. Christakis, "Let's Shake Up the Social Sciences," *New York Times*, July 21, 2013, www.nytimes.com/2013/07/21/ opinion/sunday/lets-shake-up-the-social-sciences.html; "NY Preschool Starts DNA Testing for Admission," NPR, 2012, http://m.npr.org/news/front/149804404?singlePag e=true; A. Stiny, "Warrior Gene" Defense Mounted in Santa Fe Murder Case," *Albuquerque Journal News*, 2014, www.abqjournal.com/485234/news/warrior-gene-defense-mounted-in-santa-fe-murder-case.html.

6. See, for example, Kevin M. Beaver, Matt DeLisi, Michael G. Vaughn, and J. C. Barnes, "Monoamine Oxidase A Genotype Is Associated with Gang Membership and Weapon Use," *Comprehensive Psychiatry* 51 (2010): 130–34; Jan-Emmanuel De Neve and James H. Fowler, "The MAOA Gene Predicts Credit Card Debt," *SSRN eLibrary*, 2010, http://papers.ssrn.com/sol3/papers.cfm?abstract_id=1457224; Laura Bevilacqua, Stéphane Doly, Jaakko Kaprio, Qiaoping Yuan, Roope Tikkanen, Tiina Paunio, Zhifeng

Zhou, et al., "A Population-Specific HTR2B Stop Codon Predisposes to Severe Impulsivity," *Nature* 468 (2010): 1061–66.

7. C. Camic, "Bourdieu's Cleft Sociology of Science," *Minerva: A Review of Science, Learning and Policy* 49, no.3 (2011): 275–93; A. Cambrosio and P. Keating, "The Disciplinary Stake: The Case of Chronobiology," *Social Studies of Science* 13, no. 3 (1983): 323–53; M. Fourcade, *Economists and Societies: Discipline and Profession in the United States, Britain, and France, 1890s to 1990s* (Princeton, NJ: Princeton University Press, 2009).

8. A. Clarke, "Reflections on the Reproductive Sciences in Agriculture in the UK and US, ca. 1900–2000+," *Studies in History and Philosophy of Science Part C: Studies in History and Philosophy of Biological and Biomedical Sciences* 38, no. 2 (2007): 316–39; S. Cunningham-Burley and A. Kerr, "Defining the 'Social': Towards an Understanding of Scientific and Medical Discourses on the Social Aspects of the New Human Genetics," *Sociology of Health and Illness* 21, no. 5 (1999): 647–68; W. Hong, "Domination in a Scientific Field Capital Struggle in a Chinese Isotope Lab," *Social Studies of Science* 38, no. 4 (2008): 543–70.

9. T. F. Gieryn, "Contesting Credibility Cartographically," in *Cultural Boundaries of Science: Credibility on the Line*, 27–64 (Chicago: University of Chicago Press, 1999); Aaron Panofsky, *Misbehaving Science: Controversy and the Development of Behavior Genetics* (Chicago: University of Chicago Press, 2014); Catherine Bliss, *Race Decoded: The Genomic Fight for Social Justice* (Stanford, CA: Stanford University Press, 2012).

10. S. Hilgartner, "The Dominant View of Popularization: Conceptual Problems, Political Uses," *Social Studies of Science* 20, no. 3 (1990): 519–39; Pierre Bourdieu, "The State of the Question," in *Science of Science and Reflexivity* (Chicago: University of Chicago Press, 2004).

11. Panofsky, *Misbehaving Science.*

12. K. Nathaus and H. Vollmer, "Moving Inter Disciplines: What Kind of Cooperation Are Interdisciplinary Historians and Sociologists Aiming For?," *Journal of History and Science* 1, no. 1 (2010), http://pub.uni-bielefeld.de/publication/1998719; Aaron Panofsky, "Generating Sociability to Drive Science: Patient Advocacy Organizations and Genetics Research," *Social Studies of Science* 41, no. 1 (2011): 31–57.

13. D. L. Kleinman, *Impure Cultures: University Biology and the World of Commerce* (Madison: University of Wisconsin Press, 2003); R. V. Burri, "Doing Distinctions Boundary Work and Symbolic Capital in Radiology," *Social Studies of Science* 38, no. 1 (2008): 35–62; M. H. Cooper, "Commercialization of the University and Problem Choice by Academic Biological Scientists," *Science, Technology and Human Values* 34, no. 5 (2009): 629–53; K. Moore, *Disrupting Science: Social Movements, American Scientists, and the Politics of the Military, 1945–1975* (Princeton, NJ: Princeton University Press, 2008).

14. S. Frickel and N. Gross, "A General Theory of Scientific/Intellectual Movements," *American Sociological Review* 70, no. 2 (2005): 204–32.

15. Paul Rabinow, *Making PCR : A Story of Biotechnology* (Chicago: University of Chicago Press, 1996); *French DNA : Trouble in Purgatory* (Chicago: University of Chicago Press, 1999); *Anthropos Today: Reflections on Modern Equipment* (Princeton, NJ:

Princeton University Press, 2003); S. Shapin, *The Scientific Life: A Moral History of a Late Modern Vocation* (Chicago: University of Chicago Press, 2008).

16. J. Kempner, "The Chilling Effect: How Do Researchers React to Controversy?," *PLoS Med* 5, no. 11 (2008): e222.

17. Adele Clarke, Jennifer Fishman, Jennifer R. Fosket, Laura Mamo, and Janet K. Shim, *Biomedicalization: Technoscience, Health, and Illness in the US* (Durham, NC: Duke University Press, 2010); Nikolas S. Rose, *Politics of Life Itself: Biomedicine, Power, and Subjectivity in the Twenty-First Century* (Princeton, NJ: Princeton University Press, 2007); Sara Shostak, *Exposed Science* (Berkeley: University of California Press, 2013).

18. Paul Rabinow, *Essays on the Anthropology of Reason* (Princeton, NJ: Princeton University Press, 1996).

19. Dorothy Nelkin and M. Susan Lindee, *The DNA Mystique: The Gene as a Cultural Icon* (Ann Arbor: University of Michigan Press, 2004).

20. Jerry A. Jacobs and Scott Frickel, "Interdisciplinarity: A Critical Assessment," *Annual Review of Sociology* 35 (2009): 43–65.

21. D. Stokols, S. Misra, R. P. Moser, K. L. Hall, and B. K. Taylor, "The Ecology of Team Science: Understanding Contextual Influences on Transdisciplinary Collaboration," *American Journal of Preventive Medicine* 35, no. 2 (2008): S96–S115; S. M. Fiore, "Interdisciplinarity as Teamwork How the Science of Teams Can Inform Team Science," *Small Group Research* 39, no. 3 (2008): 251–77.

22. Wesley Shrum, "Collaborationism," in *Collaboration in the New Life Sciences*, edited by John N. Parker, Niki Vermeulen, and Bart Penders, 247–58 (London and New York: Routledge, 2016).

23. S. Laberge, M. Albert, and B. D. Hodges, "Perspectives of Clinician and Biomedical Scientists on Interdisciplinary Health Research," *Canadian Medical Association Journal* 18, no. 11 (2009): 797–803; Michèle Lamont, G. Mallard, and J. Guetzkow, "Beyond Blind Faith: Overcoming the Obstacles to Interdisciplinary Evaluation," *Research Evaluation* 15, no. 1 (2006): 43–55; D. Stokols, K. L. Hall, B. K. Taylor, and R. P. Moser, "The Science of Team Science: Overview of the Field and Introduction to the Supplement," *American Journal of Preventive Medicine* 35, no. 2 (2008): S77–S89.

24. M. Albert, S. Laberge, and B. D. Hodges, "Boundary-Work in the Health Research Field: Biomedical and Clinician Scientists' Perceptions of Social Science Research," *Minerva: A Review of Science, Learning and Policy* 47, no. 2 (2009): 171–194; Pierre Bourdieu, *Pascalian Meditations* (Stanford, CA: Stanford University Press, 2000); C. Brosnan, "The Significance of Scientific Capital in UK Medical Education," *Minerva: A Review of Science, Learning and Policy* 49, no. 3 (2011): 317–32; S. Timmermans and M. Berg, *The Gold Standard: The Challenge of Evidence-Based Medicine and Standardization in Health Care* (Philadelphia: Temple University Press, 2003).

25. See A. Chettiparamb, "Interdisciplinarity: A Literature Review," in *The Higher Education Academy: Interdisciplinary Teaching and Learning Group* (Southampton, UK: Interdisciplinary Teaching and Learning Group, Subject Centre for Languages, Linguistics and Area Studies, School of Humanities, University of Southampton, 2007); P. A. Chow-White and T. Duster, "Do Health and Forensic DNA Databases Increase Racial

Disparities?," *PLoS Med* 8, no. 10 (2011): e1001100; Scott Frickel, "Mobilizing Science: Movements, Participation, and the Remaking of Knowledge," *Isis* 101, no. 4 (2010): 923–23.

26. See, for example, Ruha Benjamin, *People's Science: Bodies and Rights on the Stem Cell Frontier* (Stanford, CA: Stanford University Press, 2013); G. C. Bowker and S. L. Star, *Sorting Things Out: Classification and Its Consequences* (Cambridge, MA: MIT Press, 1999); Steven Epstein, *Inclusion: The Politics of Difference in Medical Research* (Chicago: University of Chicago Press, 2007); A. Goodman, D. Heath, and M. S. Lindee, *Genetic Nature/Culture* (Berkeley: University of California Press, 2003); Sheila Jasanoff, *Reframing Rights: Bioconstitutionalism in the Genetic Age* (Cambridge, MA: MIT Press, 2011); Laura Mamo, *Queering Reproduction* (Durham, NC: Duke University Press, 2007); Michael Montoya, *Making the Mexican Diabetic: Race, Science, and the Genetics of Inequality* (Berkeley: University of California Press, 2011); Ann Morning, *The Nature of Race* (Berkeley: University of California Press, 2011); Alondra Nelson, *The Social Life of DNA: Race, Reparations, and Reconciliation after the Genome* (Boston: Beacon Press, 2015); Osagie Obasogie and Marcy Darnovsky, *Beyond Bioethics* (Berkeley: University of California Press, 2016); Sarah Richardson, *Sex Itself: The Search for Male and Female in the Human Genome* (Chicago: University of Chicago Press, 2013); Dorothy Roberts, *Fatal Invention: How Science, Politics, and Big Business Re-create Race in the Twenty-First Century* (New Press, 2011).

27. F. S. Collins, E. D. Green, A. E. Guttmacher, and M. S. Guyer, on behalf of the US NHGRI, "A Vision for the Future of Genomics Research: A Blueprint for the Genomic Era," *Nature* 422 (2003): 835–47; Geoff Spencer, "NHGRI Funds Researchers to Evaluate Standard Measures in Genomic Studies," NIH: National Human Genome Research Institute, 2011, www.genome.gov/27546443.

28. Edwin Black, *War against the Weak: Eugenics and America's Campaign to Create a Master Race* (Dialog Press, 2008); Ladelle McWhorter, *Racism and Sexual Oppression in Anglo-America: A Genealogy* (Bloomington: Indiana University Press, 2009); R. Proctor, *Racial Hygiene: Medicine under the Nazis* (Cambridge, MA: Harvard University Press, 1988).

29. D. Nelkin and M. S. Lindee, *The DNA Mystique: The Gene as a Cultural Icon* (Ann Arbor: University of Michigan Press, 2004); D. Kevles, *In the Name of Eugenics: Genetics and the Uses of Human Heredity* (Berkeley: University of California Press, 1985); W. Provine, *The Origins of Theoretical Population Genetics* (Chicago: University of Chicago Press, 2001).

30. UNESCO, "The Race Question," July 1950, www.honestthinking.org/en/unesco/UNESCO.1950.Statement_on_Race.htm.

31. Troy Duster, *Backdoor to Eugenics* (New York: Routledge, 2003); S. Sarkar, *Genetics and Reductionism* (New York: Cambridge University Press, 1998); Panofsky, *Misbehaving Science.*

32. S. Richardson, "Race and IQ in the Postgenomic Age: The Microcephaly Case," *BioSocieties* 6 (2011): 420–46; J. Brooks and M. Ledford, "Geneticizing Disease: Implications for Racial Health Disparities," Center for American Progress, 2008, www.americanprogress.org/issues/healthcare/report/2008/01/15/3832/geneticizing-disease

-implications-for-racial-health-disparities/; R. Jordan-Young, *Brain Storm: The Flaws in the Science of Sex Differences* (Cambridge, MA: Harvard University Press, 2010); Anne Fausto-Sterling, *Sexing the Body: Gender Politics and the Construction of Sexuality* (New York: Basic Books, 2000).

33. Michael Balter, "Brain Man Makes Waves with Claims of Recent Human Evolution," *Science* 314 (2006): 1871–73.

34. "Can Your Genes Make You Murder?," NPR, 2010, www.npr.org/templates/story/story.php?storyid=128043329; B. Weiser, "Child Pornography Term Overturned Based on Judge's Genetics Theory," *New York Times*, January 29, 2011, www.nytimes.com/2011/01/29/nyregion/29ruling.html?_r=0; "NY Preschool Starts DNA Testing for Admission"; Gina Kolata, "Scientists to Seek Clues to Violence in Genome of Gunman in Newtown, Conn.," *New York Times*, December 25, 2012, www.nytimes.com/2012/12/25/science/scientists-to-seek-clues-to-violence-in-genome-of-gunman-in-newtown-conn.html.

35. A. Lippman, "Led (Astray) by Genetic Maps: The Cartography of the Human Genome and Health Care," *Social Science and Medicine* 35 (1992): 1469–76; "Prenatal Genetic Testing and Screening: Constructing Needs and Reinforcing Inequities," *American Journal of Law and Medicine* 17 (1991): 15–50; Evelyn Fox Keller, Daniel J. Kevles, and Leroy Hood, "Nature, Nurture and the Human Genome Initiative," in *The Code of Codes: Scientific and Social Issues in the Human Genome Project* (Cambridge, MA: Harvard University Press, 1992); Allison Morse, "Searching for the Holy Grail: The Human Genome Project and Its Implications," *Journal of Law and Health* 13 (1999): 219.

36. Peter Conrad, "A Mirage of Genes," *Sociology of Health and Illness* 21, no. 199: 228–41.

37. Alan V. Horwitz, "Media Portrayals and Health Inequalities: A Case Study of Characterizations of Gene X Environment Interactions," *Journal of Gerontology* 60B (2005): 48–52; Emma Kowal and G Frederick, "Race, Genetic Determinism and the Media: An Exploratory Study Media Coverage of Genetics and Indigenous Australians," *Genomics, Society and Policy* 8 (2012): 1–14.

38. I. Dar-Nimrod and S. Heine, "Genetic Essentialism: On the Deceptive Determinism of DNA," *Psychological Bulletin* 137 (2011): 800–818.

39. Y. Cheng, C. Condit, and D. Flannery, "Depiction of Gene-Environment Relationships in Online Medical Recommendations," *Genetics in Medicine* 10 (2008): 450–56.

40. E. F. Keller, "Nature and the Natural," *BioSocieties* 3 (2008): 117–24; S. Melendro-Oliver, "Shifting Concepts of Genetic Disease," *Science Studies* 17 (2004): 20–33.

41. I. De Melo-Martin, "Firing Up the Nature/Nurture Controversy: Bioethics and Genetic Determinism," *Journal of Medical Ethics* 31 (2005): 526–30; D. Heath, R. Rapp, and K. S. Taussig, "Genetic Citizenship," *A Companion to the Anthropology of Politics* (2004): 152–67; Jasanoff, *Reframing Rights*; Y. Tsai, "Geneticizing Ethnicity: A Study on the 'Taiwan Bio-Bank,'" *East Asian Science, Technology and Society* 4 (2010): 433–55; A. Clarke, J. Fishman, J. R. Fosket, L. Mamo, and Janet K. Shim, *Biomedicalization: Technoscience, Health, and Illness in the US* (Durham, NC: Duke University Press, 2010); C. Bliss,

"Defining Health Justice in the Postgenomic Era," in *Postgenomics*, edited by Sarah S. Richardson and Hallam Stevens (Durham, NC: Duke University Press, 2015).

42. Soraya De Chaderevian and Harmke Kamminga, *Molecularizing Biology and Medicine: New Practices and Alliances, 1910s–1970s* (Amsterdam: Harwood, 1998); Lily Kay, *The Molecular Vision of Life: Caltech, the Rockefeller Foundation and the New Biology* (New York: Oxford University Press, 1993); Rose, *Politics of Life Itself*; Shostak, *Exposed Science.*

43. Roberts, *Fatal Invention*; Nelson, *Social Life of DNA.*

44. Jonathan Metzl, *The Protest Psychosis: How Schizophrenia Became a Black Disease* (Boston: Beacon Press, 2009); Alondra Nelson, *Body and Soul: The Black Panther Party and the Fight against Medical Discrimination* (Minneapolis: University of Minnesota Press, 2011).

45. J. Keller, "In Genes We Trust: The Biological Component of Psychological Essentialism and Its Relationship to Mechanisms of Motivated Social Cognition," *Journal of Personality and Social Psychology* 88 (2005): 686–702.

46. Toby Jayaratne, "White and African-American Genetic Explanations for Gender, Class, and Race Differences." National Institutes of Health, 2002.

47. J. Schnittker, J. Freese, and B. Powell, "Nature, Nurture, Neither, Nor: Black-White Differences in Beliefs about the Causes and Appropriate Treatment of Mental Illness," *Social Forces* 78 (200): 1101–32.

48. Toby Epstein Jayaratne, Oscar Ybarra, Jane P. Sheldon, Tony N. Brown, Merle Feldbaum, Carla A. Pfeffer, and Elizabeth M. Petty, "White Americans' Genetic Lay Theories of Race Differences and Sexual Orientation: Their Relationship with Prejudice toward Blacks, and Gay Men and Lesbians," *Group Processes and Intergroup Relations* 9 (2006): 77–94.

49. Celeste M. Condit, R. L. Parrott, and Beth O'Grady, "Principles and Practice of Communication Processes for Genetics in Public Health," in *Genetics and Public Health in the 21st Century*, edited by Muin J. Khoury, Wylie Burke, and Elizabeth Thomson (New York: Oxford University Press, 2000); Celeste M. Condit, Alex Ferguson, Rachel Kassel, Chitra Thadhani, Holly Catherine Gooding, and Roxanne Parrott, "An Exploratory Study of the Impact of News Headlines on Genetic Determinism," *Science Communication* 22, no. 4 (2001): 379–95; C. M. Condit, R. L. Parrott, B. R. Bates, J. Bevan, and P. J. Achter, "Exploration of the Impact of Messages about Genes and Race on Lay Attitudes," *Clinical Genetics* 66 (2004): 402–8; B. R. Bates, K. Poirot, T. M. Harris, C. M. Condit, and P. J. Achter, "Evaluating Direct-to-Consumer Marketing of Race-Based Pharmacogenomics: A Focus Group Study of Public Understandings of Applied Genomic Medication," *Journal of Health Communication* 9 (2004): 541–59.

50. John Lynch, Jennifer Bevan, Paul Achter, Tina Harris, and Celeste M. Condit, "A Preliminary Study of How Multiple Exposures to Messages about Genetics Impact on Lay Attitudes towards Racial and Genetic Discrimination," *New Genetics and Society* 27 (2008): 43–56.

51. Jo C. Phelan, Bruce G. Link, and Naumi M. Feldman, "The Genomic Revolution and Beliefs about Essential Racial Differences a Backdoor to Eugenics?," *American*

Sociological Review 78 (2013): 167–91; C. Condit, A. Templeton, B. R. Bates, J. L. Bevan, and T. M. Harris, "Attitudinal Barriers to Delivery of Race-Targeted Pharmacogenomics among Informed Lay Persons," *Genetics in Medicine* 5 (2003): 385–92; C. M. Condit, R. L. Parrott, T. M. Harris, J. Lynch, and T. Dubriwny, "The Role of 'Genetics' in Popular Understandings of Race in the United States," *Public Understanding of Science* 13 (2004): 249–72; R. L. Parrott, K. J. Silk, M. R. Dillow, J. L. Krieger, T. M. Harris, and C. M. Condit, "Development and Validation of Tools to Assess Genetic Discrimination and Genetically Based Racism," *Journal of the National Medical Association* 97 (2005): 980–90.

52. Pilar Ossorio and Troy Duster, "Race and Genetics: Controversies in Biomedical, Behavioral, and Forensic Sciences," *American Psychologist* 60 (2005): 115–28; J. De Vries, M. Jallow, T. Williams, D. Kwiatkowski, M. Parker, and R. Fitzpatrick, "Investigating the Potential for Ethnic Group Harm in Collaborative Genomics Research in Africa: Is Ethnic Stigmatisation Likely?," *Social Science and Medicine* 75 (2012): 1400–1407.

53. C. E. Tygart, "Genetic Causation Attribution and Public Support of Gay Rights," *International Journal of Public Opinion Research* 12 (2000): 259–75.

54. Ibid.; Jayaratne et al., "White Americans' Genetic Lay Theories of Race Differences and Sexual Orientation."

55. Claude Steele, "Stereotype Threat and African-American Student Achievement," in *Young, Gifted, and Black: Promoting High Achievement among African-American Students*, edited by Theresa Perry, Claude Steele, and Asa Hilliard, 109–30 (Boston: Beacon Press, 2003); Matthew S. McGlone and Joshua Aronson, "Stereotype Threat, Identity Salience, and Spatial Reasoning," *Journal of Applied Developmental Psychology* 27, no. 5 (2006): 486–93.

56. Andrea Farkas Patenaude, "Pediatric Psychology Training and Genetics: What Will Twenty-First-Century Pediatric Psychologists Need to Know?," *Journal of Pediatric Psychology* 28, no. 2 (2003): 135–45.

57. Madelyn D. Freundlich, "The Case against Preadoption Genetic Testing," *Child Welfare* 77, no. 6 (1998): 663–79.

58. S. Michie, M. Bobrow, and T. M. Marteau, "Predictive Genetic Testing in Children and Adults: A Study of Emotional Impact," *Journal of Medical Genetics* 38, no. 8 (2001): 519–26; P. J. Malpas, "Predictive Genetic Testing of Children for Adult-Onset Diseases and Psychological Harm," *Journal of Medical Ethics* 34, no. 4 (2008): 275–78; Christopher H. Wade, Benjamin S. Wilfond, and Colleen M. McBride, "Effects of Genetic Risk Information on Children's Psychosocial Wellbeing: A Systematic Review of the Literature," *Genetics in Medicine* 12, no. 6 (2010): 317–26.

59. Pascal Borry, Mahsa Shabani, and Heidi Carmen Howard, "Is There a Right Time to Know? The Right Not to Know and Genetic Testing in Children," *Journal of Law, Medicine and Ethics* 42, no. 1 (2014): 19–27.

60. Even in research on adults, only one study of behavior genetic evidence on the adjudication of adult criminal behavior examined nondisease phenotypes and their relationship to institutional research uses. This study found that subjects posed as jurors who were presented with nongenetically and genetically rationalized vignettes held the greatest fear of "genetic and genetic + abuse conditions" and thus they held the greatest propensity to give harsher sentences for such individuals. Paul S. Appelbaum and

Nicholas Scurich, "Impact of Behavioral Genetic Evidence on the Adjudication of Criminal Behavior," *Journal of the American Academy of Psychiatry and the Law* Online 42, no. 1 (2014): 91–100.

61. P. Conrad, "Genetic Optimism: Framing Genes and Mental Illness in the News," *Culture, Medicine and Psychiatry* 25 (2001): 225–47; Cunningham-Burley and Kerr, "Defining the 'Social'"; A. M. Hedgecoe, *Narratives of Geneticization: Cystic Fibrosis, Diabetes and Schizophrenia* (London: University of London Press, 2000); Sara Shostak, "The Emergence of Toxicogenomics: A Case Study of Molecularization," *Social Studies of Science* 35 (2005): 367–403; S. Shostak, P. Conrad, and A. V. Horwitz, "Sequencing and Its Consequences: Path Dependence and the Relationships between Genetics and Medicalization," *American Journal of Sociology* 114 (2008): S287–S316.

62. A. M. Hedgecoe, "Terminology and the Construction of Scientific Disciplines: The Case of Pharmacogenomics," *Science, Technology and Human Values* 28 (2003): 513–37; Peter Conrad, "Uses of Expertise: Sources, Quotes, and Voice in the Reporting of Genetics in the News," *Public Understanding of Science* 8 (1999): 285–302.

63. Scott Frickel, *Chemical Consequences: Environmental Mutagens, Scientist Activism, and the Rise of Genetic Toxicology* (New Brunswick, NJ: Rutgers University Press, 2004); Shostak, *Exposed Science*.

64. Montoya, *Making the Mexican Diabetic*; R. Rapp, "Chasing Science: Children's Brains, Scientific Inquiries, and Family Labors," *Science, Technology and Human Values* 36 (2011): 662–84; S. Shostak and M. Waggoner, "Narration and Neuroscience: Encountering the Social on the 'Last Frontier of Medicine,'" in *Sociological Reflections on the Neurosciences (Advances in Medical Sociology, Vol. 13)* (Bingley, UK: Emerald Group Publishing, 2011), 51–74; Keith Wailoo, *How Cancer Crossed the Color Line* (New York: Oxford University Press, 2011).

65. Christian Torres, "Telemedicine Has More than a Remote Chance in Prisons," *Nature Medicine* 16 (2010): 496.

66. Joseph Goldstein, "Judge Rejects New York's Stop-and-Frisk Policy," *New York Times*, August 13, 2013, www.nytimes.com/2013/08/13/nyregion/stop-and-frisk-practice-violated-rights-judge-rules.html; Michèle Alexandre, *Sexploitation: Sexual Profiling and the Illusion of Gender* (New York: Routledge, 2014).

67. Troy Duster, "Selective Arrests, an Ever-Expanding DNA Forensic Database, and the Specter of an Early Twenty-First-Century Equivalent of Phrenology," in *DNA and the Criminal Justice System: The Technology of Justice*, edited by David Lazer (Cambridge, MA: MIT Press, 2004).

68. Bliss, *Race Decoded*.

69. Publications and news media were compiled through Google Scholar, Google, PubMed, and LexisNexis searches. As for other materials, such as unpublished documents, I obtained copies from researchers during interviews and observation.

70. Bliss, *Race Decoded*.

71. Robin O. Andreasen, "The Cladistic Race Concept: A Defense," *Biology and Philosophy* 19, no. 3 (2004): 425–42.

72. Catherine Bliss, "Genetic Approaches to Health Disparities," in *Advances in Medical Sociology, Vol. 16* (Bingley, UK: Emerald Group Publishing, 2015).

73. Sarah Richardson and Hallam Stevens, eds., *Postgenomics: Biology after the Genome* (Durham, NC: Duke University Press, 2014).

74. Bliss, "Defining Health Justice in the Postgenomic Era."

75. Bliss, *Race Decoded.*

76. Jantina De Vries, Muminatou Jallow, Thomas N. Williams, Dominic Kwiatkowski, Michael Parker, and Raymond Fitzpatrick, "Investigating the Potential for Ethnic Group Harm in Collaborative Genomics Research in Africa: Is Ethnic Stigmatisation Likely?," *Social Science and Medicine* 75, no. 8 (2012): 1400–1407.

77. Jasanoff, *Reframing Rights.*

78. Vural Ozdemir, David S. Rosenblatt, Louise Warnich, Sanjeeva Srivastava, Ghazi O. Tadmouri, Ramy K. Aziz, Panga Jaipal Reddy, et al., "Towards an Ecology of Collective Innovation: Human Variome Project (HVP), Rare Disease Consortium for Autosomal Loci (RaDiCAL) and Data-Enabled Life Sciences Alliance (DELSA)," *Current Pharmacogenomics and Personalized Medicine* 9, no. 4 (2011): 243–51; Ruha Benjamin, "A Lab of Their Own: Genomic Sovereignty as Postcolonial Policy," *Policy and Society* 28, no. 4 (2009): 341–55.

79. Catherine Bliss, "Translating Racial Genomics: Passages in beyond the Lab," *Qualitative Sociology* 36 (2013): 423–44.

80. Richardson and Stevens, *Postgenomics.*

81. S. Shapin, *The Scientific Life: A Moral History of a Late Modern Vocation* (Chicago: University of Chicago Press, 2008).

Chapter 1: Genes and Their Environments

1. Evelyn Fox Keller, *The Mirage of a Space between Nature and Nurture* (Durham, NC: Duke University Press, 2010).

2. Ibid.

3. According to Keller, Shakespeare's mention in *The Tempest* did not oppose the terms or conjoin them in a way that presumed that they were somehow disjointed. John Locke, who coined the term "the blank slate," also did not oppose nature to nurture. Rather, he encouraged parents and teachers to take heed of children's inborn dispositions when educating them.

4. Whereas Darwin's formulations of heredity hinted at the modern formulation of nature versus nurture, the phrase was only first explicitly stated in Galton's *English Men of Science: Their Nature and Nurture*; Sir Francis Galton, *English Men of Science: Their Nature and Nurture* (Cambridge: Macmillan, 1874).

5. Sir Francis Galton, *Hereditary Genius: An Inquiry into Its Laws and Consequences* (London: Macmillan, 1869).

6. Eugenics, literally meaning good birth or good race, is "the selection of desired heritable characteristics in order to improve future generations"; Kara Rodgers, *New Thinking about Genetics* (New York: Rosen, 2010). A broad idea of eugenics has

been around since Plato "recommended a state-run program of mating intended to strengthen the guardian class in his *Republic*"; Kevles, *In the Name of Eugenics*.

7. Galton, *Hereditary Genius*.

8. Wendy Kline, *Building a Better Race: Gender, Sexuality, and Eugenics from the Turn of the Century to the Baby Boom* (Berkeley: University of California Press, 2005).

9. Thorndike was a staunch eugenicist who said, "Selective breeding can alter man's capacity to learn, to keep sane, to cherish justice or to be happy. There is no more certain and economical a way to improve man's environment as to improve his nature." He would go on to serve as president of the American Psychological Association and to launch the subfield of educational psychology that is today associated with the "Law of Effect and Operant Conditioning"; Richard Lynn, *Eugenics: A Reassessment (Westport, CT: Praeger, 2001)*.

10. Lea Winerman, "A Second Look at Twin Studies," *Monitor on Psychology* 35 (2004): 46, www.apa.org/monitor/apr04/second.aspx.

11. Keller, *Mirage of a Space between Nature and Nurture*.

12. Lewis M. Terman et al., "The Stanford Revision and Extension of the Binet-Simon Scale for Measuring Intelligence," *Journal of Educational Psychology* 6 (1915).

13. Ibid., 7.

14. IQ tests have been critiqued for not considering the extremely varied effects on individuals' "intellect," including structural disadvantages related to class, race and gender; Steve Connor, "IQ Tests Are 'Fundamentally Flawed' and Using Them Alone to Measure Intelligence Is a 'Fallacy,' Study Finds," *Independent*, 2012, www.independent.co .uk/news/science/iq-tests-are-fundamentally-flawed-and-using-them-alone-to-measure -intelligence-is-a-fallacy-study-8425911.html.

15. Positive eugenics is the encouragement of people of good health to reproduce together, and negative eugenics is working toward ending certain diseases or disabilities by preventing certain people from reproducing; Alexandra Stern, "Making Better Babies: Public Health and Race Betterment in Indiana, 1920–1935," American Journal of Public Health 92 (2002): 742–52, www.ncbi.nlm.nih.gov/pmc/articles/PMC3222231/. Negative eugenics included the forced sterilization of people deemed unfit to reproduce—at the time, "poor, mentally insane, feeble-minded, idiots, drunken," and others; Stern, "Making Better Babies."

16. Twin studies have become the oldest and most favored research method for the field of behavior genetics; Lea Winerman, "Behavioral Genetics—A Second Look at Twin Studies," American Psychological Association, 2004, www.apa.org/monitor/ apr04/second.aspx.

17. Alexandra Stern, *Eugenic Nation: Faults and Frontiers of Better Breeding in Modern America* (Berkeley: University of California Press, 2005).

18. Ibid., 47. As Keller points out, medical genetic diagnosis couldn't really say anything about causal pathways, since it only tells us about genetic sequences and their anomalies.

19. Ibid.

20. Panofsky, *Misbehaving Science*.

21. Tim Spector, *Identically Different: Why We Can Change Our Genes* (New York: Overlook Press, 2014).

22. Winerman, "Behavioral Genetics."

23. Panofsky, *Misbehaving Science.*

24. Ibid.

25. Jenny Reardon, *Race to the Finish: Identity and Governance in an Age of Genomics* (Princeton, NJ: Princeton University Press, 2005).

26. E. O. Wilson, *Sociobiology: The New Synthesis* (Cambridge, MA: Harvard University Press, 2005).

27. E. O. Wilson, *On Human Nature* (Cambridge, MA: Harvard University Press, 1978).

28. David Inglis, John Bone, and Rhoda Wilkie, *Nature: The Nature of Human Nature* (New York: Taylor & Francis, 2005).

29. See Robert L. Trivers, "The Evolution of Reciprocal Altruism," *Quarterly Review of Biology* 46 (1971): 35–57; Michael Ruse and Joseph Travis, *Evolution: The First Four Billion Years* (Cambridge, MA: Harvard University Press, 2009); Donald Symons, *The Evolution of Sexuality* (New York: Oxford University Press, 1979); John Tooby and Leda Cosmides, "The Psychological Foundations of Culture," in *The Adapted Mind: Evolutionary Psychology and the Generation of Culture* (New York: Oxford University Press, 1992).

30. Philip Kitcher, *Vaulting Ambition: Sociobiology and the Quest for Human Nature* (Cambridge, MA: MIT Press, 1987).

31. Richard Lewontin, "Sociobiology as an Adaptationist Program," *Behavioral Science* 24 (1979): 5–14.

32. David J. Buller, *Adapting Minds: Evolutionary Psychology and the Persistent Quest for Human Nature* (Cambridge, MA: MIT Press, 2006); Leda Cosmides and John Tooby, "From Evolution to Behavior: Evolutionary Psychology as the Missing Link," in *The Latest on the Best: Essays on Evolution and Optimality*, by John Dupre (Cambridge, MA: MIT Press, 1987).

33. Sociobiologists have been similarly critiqued for rationalizing violence toward women and children, such as rape and sexual abuse; J. D. Anker, *Systems and Theories in Psychology* (New York: McGraw-Hill, 1987).

34. Yudell also notes that Rushton served as a "bridge between eugenics and the mainstream realm of sociobiology"; Michael Yudell, *Race Unmasked: Biology and Race in the Twentieth Century* (New York: Columbia University Press, 2014), 188.

35. David M. Buss, *The Handbook of Evolutionary Psychology* (Hoboken, NJ: Wiley, 2005).

36. E. A. Smith, M. Borgerhoff Mulder, and K. Hill, "Controversies in the Evolutionary Social Sciences: A Guide for the Perplexed," *Trends in Ecology and Evolution* 16, no. 3 (2001): 128–35.

37. Using data compiled in the National Longitudinal Study of Youth (NLSY), the authors analyzed IQ scores of all participants in the NLSY and evaluated participants for social and economic outcomes; Richard J. Herrnstein and Charles A. Murray, *The*

Bell Curve: Intelligence and Class Structure in American Life (New York: Free Press, 1994). This methodology has been since critiqued for manipulating statistics, mis-citing sources, and "sloppy reasoning"; Nicholas Lemann, "The Bell Curve Flattened," *Slate*, 1997, www.slate.com/articles/briefing/articles/1997/01/the_bell_curve_flattened.html.

38. Panofsky, *Misbehaving Science*, 1.

39. Herrnstein and Murray, *Bell Curve*.

40. More specifically, the book argued that "US class structure can mostly be attributed to inequalities in individual intelligence as measured by IQ, that IQ is mostly an innate capacity of individuals under genetic control, and therefore differences in education and upbringing are not responsible for social inequalities"; Panofsky, *Misbehaving Science*. Further, they claimed that "genetic differences largely explain the lack of black and Latino success relative to white and Asian, though environment plays some role"; ibid.

41. Herrnstein and Murray, *Bell Curve*, 548–49.

42. Jason DeParle, "Daring Research or 'Social Science Pornography'?: Charles Murray," *New York Times*, October 9, 1994, www.nytimes.com/1994/10/09/magazine/daring -research-or-social-science-pornography-charles-murray.html?pagewanted=all; Michael Land, "In Response to 'Race, Genes and I.Q.,'" *New Republic*, 1994, https://newrepublic .com/article/120890/responses-new-republics-bell-curve-excerpt.

43. William J. Matthews, "A Review of *The Bell Curve*: Bad Science Makes for Bad Conclusions," *David Boles Blogs*, March 23, 1998, http://bolesblogs.com/1998/03/23/a -review-of-the-bell-curve-bad-science-makes-for-bad-conclusions/.

44. Panofsky, *Misbehaving Science*, 16.

45. Ibid.

46. John P. Jackson and Nadine M. Weidman, *Race, Racism, and Science: Social Impact and Interaction* (Santa Barbara, CA: ABC-CLIO, 2006), 221.

47. Panofsky writes that many scientists rallied around Rushton and Whitney's academic freedom, after which "Whitney subsequently dove head first into white supremacist politics, contributing the forward to white supremacist David Duke's autobiography and writing for the extreme right wing magazine *American Renaissance*, before he died in 2002"; Panofsky, *Misbehaving Science*, 4–5.

48. Jim Naureckas, "Racism Resurgent: How Media Let *The Bell Curve*'s Pseudo-Science Define the Agenda on Race," *Fairness and Accuracy in Reporting*, 1995, http:// fair.org/extra-online-articles/racism-resurgent/.

49. Ibid.

50. As one historiographer writes, "Evolutionary psychology did not seek to replace sociobiology, but instead it strengthened and expanded upon the strongest sociobiological theses, incorporating these new and improved ideas into its infrastructure"; Melissa Seltin, "The Evolution of Evolutionary Psychology: From Sociobiology to Evolutionary Psychology," *Personality Research*, 1998, www.personalityresearch.org/papers/seltin. html.

51. Buss, *Handbook of Evolutionary Psychology*.

52. Tooby and Cosmides, "Psychological Foundations of Culture."

53. Ibid.

54. Robert C. Richardson, *Evolutionary Psychology as Maladapted Psychology (Life and Mind: Philosophical Issues in Biology and Psychology)* (Cambridge, MA: MIT Press, 2010).

55. Henry Plotkin, *Evolutionary Thought in Psychology: A Brief History* (Malden, MA: Blackwell, 2004), 149.

56. Kenan Malik, "Darwinian Fallacies," *Prospect*, 1998, www.kenanmalik.com/essays/fallacy.html.

57. Ibid.

58. Satoshi Kanazawa, "Why Are Black Women Less Physically Attractive than Other Women?," *Psychology Today, The Scientific Fundamentalist* (blog), 2011; *Why Men Gamble and Women Buy Shoes: How Evolution Shaped the Way We Behave* (Oxford: Oxford University Press, 2006). Kanazawa's other incendiary statements include claiming that Americans won't win wars in the Middle East because they don't hate their enemies enough. He writes:

> Imagine that, on September 11, 2001, when the Twin Towers came down, the President of the United States was not George W. Bush, but Ann Coulter. What would have happened then? On September 12, President Coulter would have ordered the US military forces to drop 35 nuclear bombs throughout the Middle East, killing all of our actual and potential enemy combatants, and their wives and children. On September 13, the war would have been over and won, without a single American life lost.

Satoshi Kanazawa, "Why We Are Losing This War: All You Need Is Hate," *Psychology Today, The Scientific Fundamentalist*, 2008, www.psychologytoday.com/blog/the-scientific-fundamentalist/200803/why-we-are-losing-war.

59. Kanazawa, "Why Are Black Women Less Physically Attractive than Other Women?"

60. Natasha Lennard, "'Why Are Black Women Less Attractive?' Asks *Psychology Today*," *Salon*, 2011, www.salon.com/2011/05/17/psychology_today_racist_black_women_attractive/.

61. Kaja Perina, "An Apology from *Psychology Today*: Statement from the Editor," *Psychology Today, Brainstorm* (blog), 2011, www.psychologytoday.com/blog/brainstorm/201105/apology-psychology-today.

62. Ibid.

63. Ujala Sehgal, "Professor Faces Removal over 'Black Women Less Attractive' Claim," *The Wire*, 2011, www.thewire.com/national/2011/05/professor-may-lose-job-over-racist-article-black-women/38008/.

64. Alex Beaujean et al., "Sinned Against, Not Sinning," *Times Higher Education*, 2011, www.timeshighereducation.co.uk/comment/letters/sinned-against-not-sinning/416527.article.

65. Panofsky, *Misbehaving Science*, 226.

66. Richardson, "Race and IQ in the Postgenomic Age."

67. Ibid.

68. Ibid. One example Richardson presents is with genetic anthropologists Harpending and Cochran's statements from *The 10,000 Year Explosion*, in which they write, "The obvious between-population differences that we knew of a few years ago were only the

tip of the iceberg," and later, "The most interesting genetic changes are surely those that change minds rather than bodies"; Henry Harpending and Gregory Cochran, *The 10,000 Year Explosion: How Civilization Accelerated Human Evolution* (New York: Basic Books, 2009). Richardson traces their out-of-Africa intelligence evolution theory which maintains that intelligence evolved after humans left Africa.

69. The Society for Neuroeconomics defines the field as "a nascent field that represents the confluence of economics, psychology and neuroscience in the study of human decision making"; "About the Society," *Neuroeconomics*, 2015, http://neuroeconomics .org/about-the-society/.

70. Nick Wilkinson and Matthias Klaes, *An Introduction to Behavioral Economics* (London: Palgrave Macmillan, 2012), 54.

71. Paul W. Glimcher et al., "Introduction: A Brief History of Neuroeconomics," in *Neuroeconomics: Decision Making and the Brain* (London: Elsevier, 2009).

72. Glimcher et al., "Introduction: A Brief History of Neuroeconomics," xxiii.

73. Two major critics of the emerging field are Glenn W. Harrison and Emanuel Donchin, who have argued that neuroeconomics "misunderstand and underestimate traditional economic models"; Ariel Rubinstein, "Discussion of 'Behavioral Economics,' 'Behavioral Economics' (Collin Camerer) and 'Incentives and Self-Control' (Matthew Rabin)," in *Advances in Economics and Econometrics Theory and Applications, Ninth World Congress* (Cambridge: Cambridge University Press, 2006).

74. Faruk Gul and Wolfgang Pesendorfer, "The Case for Mindless Economics," Princeton University, 2005, www.princeton.edu/~pesendor/mindless.pdf.

75. See, for example, Drew Westen et al., "Neural Bases of Motivated Reasoning: An fMRI Study of Emotional Constraints on Partisan Political Judgment in the 2004 US Presidential Election," *Journal of Cognitive Neuroscience* 18 (2006): 1947–58.

76. Stuart Henry and Dena Plemmons, "Neuroscience, Neuropolitics and Neuroethics: The Complex Case of Crime, Deception and fMRI," *Science and Engineering Ethics* 19 (2012): 573–91, www.academia.edu/3156669/Neuroscience_Neuropolitics_and_ Neuroethics_The_Complex_Case_of_Crime_Deception_and_fMRI.

77. Nicole Rafter, *The Criminal Brain: Understanding Biological Theories of Crime* (New York: New York University Press, 2008), 199.

78. Peter Monaghan, "Biocriminology: Genetic Links in a Criminal Chain," Center for Genetics and Society, 2009, www.geneticsandsociety.org/article.php?id=4713.

79. Ibid.

80. Robert Pool, "The Rise of Biocriminology," *Florida State University: Research in Review* (Fall/Winter 2009), www.rinr.fsu.edu/issues/2009fallwinter/feature02_d.asp.

81. Cesare Lombroso-Ferrero, *Criminal Man According to the Classification of Cesare Lombroso Briefly Summarized by His Daughter Gina Lombroso-Ferrero with an Introduction by Cesare Lombroso* (Montclair: Patterson, 1911); Greta Olson, "Criminalized Bodies in Literature and Biocriminology," Academia.edu, 2007, 239–59, www.academia .edu/7113070/Criminalized_Bodies_in_Literature_and_Biocriminology.

82. Olson, "Criminalized Bodies in Literature and Biocriminology."

83. Rafter, *Criminal Brain*, 199.

84. Ibid.

85. In 1964 British psychologist Hans Eysenck published *Crime and Personality*, a genetic and neurophysiological explanation of criminality; ibid., 199. As Rafter explains, the crux of Eysenck's theory lay in classical or Pavlovian conditioning: "He portrayed criminals as extraverts, sensation seekers who engage in risky and criminal activities to compensate for the sluggish nervous systems they inherit"; ibid., 200. It was this theory that thrust trends back in the direction of biological explanations and inspired later fields such as neurocriminology; ibid., 201.

86. Ibid., 219, 214.

87. Pool, "Rise of Biocriminology."

88. Bernice A. Pescosolido, "Taking 'the Promise' Seriously: Medical Sociology's Role in Health, Illness, and Healing in a Time of Change," in *Handbook of the Sociology of Health, Illness, and Healing: A Blueprint for the 21st Century*, edited by Bernice A. Pescosolido, Jack K. Martin, Jane D. McLeod, and Anne Rogers, 3–20 (New York: Springer, 2010).

89. Bliss, *Race Decoded*.

90. Francis S. Collins, *The Language of God: A Scientist Presents Evidence for Belief* (New York: Simon & Schuster, 2006).

91. Emily Chang, "In China, DNA Tests on Kids ID Genetic Gifts, Careers," CNN, 2009, http://edition.cnn.com/2009/WORLD/asiapcf/08/03/china.dna.children.ability/.

92. David Cyranoski and Sara Reardon, "Chinese Scientists Genetically Modify Human Embryos," *Nature*, 2015, www.nature.com/doifinder/10.1038/nature.2015.17378.

93. Keller, *Mirage of a Space between Nature and Nurture*.

94. Ibid.

95. Bliss, *Race Decoded*.

96. Rabinow, "Artificiality and Enlightenment."

97. Kaushik Sunder Rajan, *Biocapital: The Constitution of Postgenomic Life* (Durham, NC: Duke University Press, 2005), 23.

98. Ian Hacking, "Genetics, Biosocial Groups and the Future of Identity," *Daedalus* 135 (Fall 2006): 81–95.

99. Rose, *Politics of Life Itself*.

100. Spector, *Identically Different*.

101. Also see Comaroff's coverage of new racial politics brought on by the global HIV/AIDS epidemic; Jean Cormaroff, "Beyond Bare Life: AIDS, (Bio)Politics, and the Neoliberal Order," *Public Culture Winter* 19, no. 1 (2007): 197–219.

102. Nelson, *Social Life of DNA*.

103. In *The Genome Factor*, social genomicists Dalton Conley and Jason Fletcher carefully tackle these issues, elaborating a number of policy landmines; *The Genome Factor: What the Social Genomics Revolution Reveals about Ourselves, Our History, Our Future* (Princeton, NJ: Princeton University Press, 2017).

104. Rajan calls this a message of "structural messianism," which is intrinsically entwined with the motivation for people to take their DNA into their own hands and manage their own health. Thus, liberal discourses of empowerment become neoliberal discourses of privatization and individualization. The new genomics only has to provide (i.e., sell) tests and the rest is up to you; Rajan, *Biocapital*.

Chapter 2: Science without Borders

1. Dalton Conley, "Learning to Love Animal (Models) (or) How (Not) to Study Genes as a Social Scientists," in *Handbook of the Sociology of Health, Illness, and Healing,* edited by Bernice Pescosolido, Jack Martin, Jane McLeod, Anne Rogers, 527–42 (New York: Springer, 2011).

2. Interview 10. All interviews conducted by the author.

3. "Add Health," National Longitudinal Study of Adolescent Health, retrieved November 6, 2015, www.cpc.unc.edu/projects/addhealth.

4. "Wave III—Add Health," National Longitudinal Study of Adolescent Health, retrieved November 6, 2015, www.cpc.unc.edu/projects/addhealth/design/wave3.

5. "Wave IV—Add Health," National Longitudinal Study of Adolescent Health, retrieved November 6, 2015, www.cpc.unc.edu/projects/addhealth/design/wave4.

6. "Purpose of Wisconsin Longitudinal Study," Wisconsin Longitudinal Study. www.ssc.wisc.edu/wlsresearch/about/.

7. "About," UK Biobank, 2015, www.ukbiobank.ac.uk/about-biobank-uk/.

8. "Home—Genetics Data," Health and Retirement Study, 2015, http://hrsonline.isr.umich.edu/gwas.

9. Avshalom Caspi et al., "Role of Genotype in the Cycle of Violence in Maltreated Children," *Science 297, no. 5582 (2002): 851–54.*

10. Avshalom Caspi et al., "Influence of Life Stress on Depression: Moderation by a Polymorphism in the 5-HTT Gene," *Science* 301, no. 5631 (2003): 386–89.

11. "Career Development K Awards," NIH National Institute of Child Health and Human Development, 2015, www.nichd.nih.gov/training/extramural/Pages/career.aspx.

12. "Award Criteria and Recipients: Early Achievement," Population Association of America, 2013, www.populationassociation.org/sidebar/annual-meeting/awards/early-achievement/.

13. Jan-Emmanuel De Neve, Slava Mikhaylov, Christopher T. Dawes, Nicholas A. Christakis, and James H. Fowler, "Born to Lead? A Twin Design and Genetic Association Study of Leadership Role Occupancy," *Leadership Quarterly* 24, no. 1 (2013): 45–60.

14. James H. Fowler, Laura A. Baker, and Christopher T. Dawes, "Genetic Variation in Political Participation," *American Political Science Review* 102, no. 2 (2008): 233–48.

15. Rose McDermott, Dustin Tingley, Jonathan Cowden, Giovanni Frazzetto, and Dominic D. P. Johnson, "Monoamine Oxidase A Gene (MAOA) Predicts Behavioral Aggression Following Provocation," *Proceedings of the National Academy of Sciences of the United States of America* 106, no. 7 (2009): 2118–23.

16. See Jonathan Daw and Guang Guo, "The Influence of Three Genes on Whether Adolescents Use Contraception, USA 1994–2002," *Population Studies* 65, no. 3 (2011): 253–71; Guang Guo, Michael E. Roettger, and Tianji Cai, "The Integration of Genetic Propensities into Social-Control Models of Delinquency and Violence among Male Youths," *American Sociological Review* 73, no. 4 (2008): 543–68; Guang Guo and Yuying Tong, "Age at First Sexual Intercourse, Genes, and Social Context: Evidence from Twins and the Dopamine D4 Receptor Gene," *Demography* 43, no. 4 (2006): 747–69.

17. Jeremy Freese and Sara Shostak, "Genetics and Social Inquiry," *Annual Review of Sociology* 35, no. 1 (2009): 107–28.

18. Peter Hatemi and Rose McDermott, *Man Is by Nature a Political Animal* (Chicago: University of Chicago Press, 2011).

19. James H. Fowler, Laura A. Baker, and Christopher T. Dawes, "Genetic Variation in Political Participation," *American Political Science Review* 102, no. 2 (2008): 233–48.

20. Guo, Roettger, and Cai, "Integration of Genetic Propensities into Social-Control Models of Delinquency and Violence among Male Youths."

21. Ariel Knafo, Israel Salomon, Darvasi Ariel, Bachner-Melman Rachel, Uzefovsky Florina, et al., "Individual Differences in Allocation of Funds in the Dictator Game Associated with Length of the Arginine Vasopressin 1a Receptor (AVPR1a) RS3 Promoter Region and Correlation between RS3 Length and Hippocampal mRNA," *Genes, Brain and Behavior* 7, no. 3(2008): 266–75.

22. Matt Bradshaw and Christopher G. Ellison, "The Nature-Nurture Debate Is Over, and Both Sides Lost! Implications for Understanding Gender Differences in Religiosity," *Journal for the Scientific Study of Religion* 48, no. 2 (2009): 241–51.

23. Daniel J. Benjamin et al., "The Promises and Pitfalls of Genoeconomics," *Annual Review of Economics* 4, no. 1 (2012): 627–62.

24. "Welcome," Philipp Koellinger, 2015, www.philipp-koellinger.com.

25. Ibid. See also "Genoeconomics and Entrepreneurship Research," Erasmus Centre for Entrepreneurship Research, 2010, www.erim.eur.nl/centres/entrepreneurship/news/featuring/detail/2181-genoeconomics-and-entrepreneurship-research/.

26. "Events," Social Science Genetic Association Consortium, 2011, http://ssgac.org/Events.php.

27. "Conference Schedule," Integrating Genetics and the Social Science, University of Colorado, 2010, www.colorado.edu/ibs/cupc/conferences/IGSS_2010/schedule.html.

28. "Welcome to the Social Science Genetic Association Consortium," Social Science Genetic Association Consortium, 2011, www.ssgac.org/Home.php.

29. Jason D. Boardman and Jason M. Fletcher, "Introduction to the Special Issue on Integrating Genetics and the Social Sciences," *Biodemography and Social Biology* 57, no. 1 (2011): 1–2.

30. Jonathan P. Beauchamp et al., "Molecular Genetics and Economics," *Journal of Economic Perspectives : A Journal of the American Economic Association* 25, no. 4 (2011): 57–82.

31. "Welcome to the Social Science Genetic Association Consortium (SSGAC)," Social Science Genetic Association Consortium, 2014, www.ssgac.org/Home.php.

32. "IGSS," Integrating Genetics and the Social Science, University of Colorado, 2010, www.colorado.edu/ibs/cupc/conferences/IGSS_2010/schedule.html.

33. "Prestigious ERC Consolidator Grant for Melinda Mills," University of Groeningen, 2014, www.rug.nl/news/2014/01/0108-erc-consolidator-grant-melinda-mills?lang=en.

34. "ERC Consolidator Grant for Phillip Koellinger," Amsterdam Business School, 2015, http://abs.uva.nl/news-events/content/2015/03/erc-consolidator-grant-for-philipp-koellinger.html.

35. "Integrative Graduate Education and Research Traineeship Program (IGERT)," National Science Foundation, 2013, www.nsf.gov/funding/pgm_summ.jsp?pims_id=12759&org=NSF.

36. "Welcome to the Erasmus University Rotterdam Institute for Behavior and Biology (EURIBEB)," Erasmus School of Economics, 2015, www.eur.nl/ese/english/departments/department_of_applied_economics/affiliated_institutes/euribeb/.

37. "Sociogenome," Sociogenome, University of Oxford, 2015, http://sociogenome.com.

38. Koellinger, "Welcome."

39. Ibid.

40. "Phenotypes," Social Science Genetic Association Consortium, 2011, http://ssgac.org/Phenotypes.php.

41. Ibid.

42. "Welcome to the Broad Institute," Broad Institute, 2015, www.broadinstitute.org.

43. For more on rhizome structures of science, see Bruno Latour, "Technology as a Society Made Durable," in *A Sociology of Monsters* (New York: Routledge, 1999).

44. Kelly Moore and Daniel Lee Kleinman, "Science and Neoliberal Globalization: A Political Sociological Approach," *Theory and Society* 40, no. 5 (2011): 505–32.

45. Daniel Lee Kleinman and Steven P. Vallas, "Science, Capitalism, and the Rise of the 'Knowledge Worker': The Changing Structure of Knowledge Production in the United States," *Theory and Society* 30 (2001): 451–92.

46. Bliss, *Race Decoded*.

47. "Interdisciplinary Research," Interdisciplinary Funding Opportunities, NIH, 2015, https://commonfund.nih.gov/interdisciplinary/grants; "Synergy Grants," European Research Council, 2014, https://erc.europa.eu/synergy-grants; "Research Environment," Medical Research Council, 2015, www.mrc.ac.uk/research/strategy/aim-4/objective-15/; Mary Bullock, "NAS Releases New Report: Facilitating Interdisciplinary Research," American Psychological Association, 2005, www.apa.org/science/about/psa/2005/05/nasa.aspx.

48. See Camille Gamboa, "How You Can Help Prevent a 45% Cut from NSF's Social, Behavioral, and Economic Research Budget," *SAGE Connection Insight*, 2015, http://connection.sagepub.com/blog/industry-news/2015/05/21/how-you-can-help-prevent-a-45-cut-from-nsfs-social-behavioral-economic-research-budget/; "House Advances Bills to Cut Social Science Funding," Consortium of Social Science Associations, 2015, www.cossa.org/2015/06/02/house-advances-bills-to-cut-social-science-funding/.

49. "Current and Former Scholars—Brown," William T. Grant Foundation, 2015, http://wtgrantfoundation.org/Current-and-Former-William-T-Grant-Scholars. Also see "New Research Grantees Will Focus on Reducing Inequality and the Use of Research Evidence," William T. Grant Foundation, 2015, http://blog.wtgrantfoundation.org/post/131173508252/new-research-grantees-will-focus-on-reducing.

50. "2014 Research Prize—Jacobs Foundation," Jacobs Foundation, 2014, http://jacobsfoundation.org/awards/2014-research-prize/.

51. "2010 Research Prize," Jacobs Foundation, 2010, http://jacobsfoundation.org/awards/research-prize-2010/.

52. "Research Funding," Jacobs Foundation, 2015, http://jacobsfoundation.org/what-we-do/research-funding/.

53. "Bios of 2013 RSI and Mini-Med School Participants," Rand Summer Institute, Rand Corporation 2013, www.rand.org/labor/aging/rsi/2013/bios.html#benjamin.

54. Ibid.

55. For further resources on the sociology of fields and disciplines, see Mathieu Albert, Suzanne Laberge, and Brian D. Hodges, "Boundary-Work in the Health Research Field: Biomedical and Clinician Scientists' Perceptions of Social Science Research," *Minerva: A Review of Science, Learning and Policy* 47, no. 2(2009): 171–94; Bourdieu, "State of the Question"; Michele Lamont, *How Professors Think: Inside the Curious World of Academic Judgment* (Cambridge, MA: Harvard University Press, 2009); K. Moore, *Disrupting Science: Social Movements, American Scientists, and the Politics of the Military, 1945–1975* (Princeton, NJ: Princeton University Press, 2008); Andrew Delano Abbott, *Chaos of Disciplines* (Chicago: University of Chicago Press, 2001); John Levi Martin, "What Is Field Theory?," *American Journal of Sociology* 109, no. 1 (2003): 1–49; Walter W. Powell and Paul J. DiMaggio, eds., *The New Institutionalism in Organizational Analysis* (Chicago: University of Chicago Press, 1991).

56. Catherine Bliss, Field notes (Integrating Genetics and Social Sciences, 2012); Field notes (Social Science Genetic Association Consortium, 2013); Field notes (Integrating Genetics and Social Sciences, 2013).

57. Ibid.

58. Actually, this statistic is not so different from the overall figure found for longstanding science, technology, engineering, and mathematics (STEM) fields, which hovers around that mark due to women's dismal underrepresentation in engineering and math, but it is completely out of sync with the typical breakdown in the staid and steady social and life sciences. In social science writ large, women are overrepresented at a whopping 61 percent (the figure being the same for the field of sociology specifically). Though economics, other social science, and various researcher positions suffer from gender gaps (economics being the worst of them with less than 33 percent representation of women), most social science fields and positions hover near the expected 50–50 ratio. Likewise, in the life sciences, specifically in biology and medicine, women are similarly represented around the 50 percent mark.

59. Interview 239.

60. Interview 216.

61. Luke Timmerman, "DNA Sequencing Market Will Exceed $20 Billion, Says Illumina CEO Jay Flatley," *Forbes*, 2015, www.forbes.com/sites/luketimmerman/2015/04/29/qa-with-jay-flatley-ceo-of-illumina-the-genomics-company-pursuing-a-20b-market/.

62. Jerry A. Jacobs and Scott Frickel, "Interdisciplinarity: A Critical Assessment," *Annual Review of Sociology* 35 (2009): 43–65.

Chapter 3: Toward the "Deeper Descriptions"

1. Interview 1.

2. Interview 3.

3. Rietveld et al., "GWAS of 126,559 Individuals Identifies Genetic Variants Associated with Educational Attainment."

4. Interview 11.

5. Rietveld et al., "GWAS of 126,559 Individuals Identifies Genetic Variants Associated with Educational Attainment."

6. Ibid.

7. Interview 31.

8. Interview 7.

9. Interview 8.

10. Fowler and Dawes, "Two Genes Predict Voter Turnout."

11. Interview 47.

12. Fowler and Dawes, "Two Genes Predict Voter Turnout," 583.

13. Peter K. Hatemi, Christopher T. Dawes, Amanda Frost-Keller, Jaime E. Settle, and Brad Verhulst, "Integrating Social Science and Genetics: News from the Political Front," *Biodemography and Social Biology* 57, no. 1 (2011): 76.

14. Interview 8.

15. Interview 54.

16. Interview 53.

17. Daniel Benjamin, Christopher Chabris, Edward Glaeser, Vilmundur Gudnason, Tamara B. Harris, David Laibson, Lenore Launer, and Shaun Purcell, "Genoeconomics," Workshop on Collecting and Utilizing Biological Indicators and Genetic Information in Social Science Surveys, 2006, 17.

18. Ibid., 17.

19. Amal Harrati, "Characterizing the Genetic Influences on Risk Aversion," *Biodemography and Social Biology* 60, no. 2 (2014): 185.

20. Ibid.

21. Maja Van der Velden, "What's Love Got to Do with IT? On Ethics and Accountability in Telling Technology Stories," in *Cultural Attitudes Towards Technology and Communication*, edited by Fay Sudweeks, Herbert Hrachovec, and Charles Ess (Murdoch University, 2008).

22. Daniel J. Benjamin, David Cesarini, Matthijs J. H. M. van der Loos, Christopher T. Dawes, Philipp D. Koellinger, Patrik K. E. Magnusson, Christopher F. Chabris, et al., "The Genetic Architecture of Economic and Political Preferences," *Proceedings of the National Academy of Sciences* 109, no. 21 (2012): 8026–31.

23. Jonathan Daw, Michael Shanahan, Kathleen Mullan Harris, et al., "Genetic Sensitivity to Peer Behaviors: 5HTTLPR, Smoking, and Alcohol Consumption," *Journal of Health and Social Behavior* 54, no. 1 (2013): 119.

24. Interview 46.

25. George L. Wehby and Stephanie Von Hinke Kessler Scholder, "Genetic Instrumental Variable Studies of Effects of Prenatal Risk Factors," *Biodemography and Social Biology* 59, no. 4 (2013): 4.

26. Ibid.

27. Interview 28.

28. See the following chapter for an in-depth analysis of scientist motivations.

29. Guang Guo and Yuying Tong, "Age at First Sexual Intercourse, Genes, and Social Context: Evidence from Twins and the Dopamine D4 Receptor Gene," *Demography* 43, no. 4 (2006): 747–69.

30. Ibid., 750.

31. Ibid.

32. See Conley and Fletcher's *The Genome Factor* for a more technically detailed analysis of the field's methods and their limitations.

33. Kevin M. Beaver, "The Intersection of Genes, the Environment, and Crime and Delinquency: A Longitudinal Study of Offending," U.S. Department of Justice, 2006, 41–42.

34. Jason D. Boardman, Kari B. Alexander, and Michael C. Stallings, "Stressful Life Events and Depression among Adolescent Twin Pairs," *Biodemography and Social Biology* 57, no. 1 (2011): 53.

35. Benjamin et al., "Genetic Architecture of Economic and Political Preferences."

36. Fowler and Dawes, "Two Genes Predict Voter Turnout."

37. Harrati, "Characterizing the Genetic Influences on Risk Aversion."

38. Some researchers critiqued GWAS for its weak explanatory power and attested to retreating from GWAS in favor of candidate gene studies. One went so far as to eschew GWAS, candidate gene studies, and twins studies for adoptee studies. He said:

I am kind of frustrated with the single gene studies or even like the two or three gene studies; I just don't think they explain much, which is what led me to the adoptee stuff in the first place, that, okay, this isn't very scientifically rigorous because we are not observing genes directly, but at least we have some credible measure of the aggregate importance of genes that implicitly incorporates all the complex interactions between them. (Interview 49)

He maintained that heritability could only be established by comparing variance between biological and adopted children.

39. See, for example, American Sociological Association, "Sociological Research Shows Combined Impact of Genetics, Social Factors on Delinquency," *E! Science News*, 2008, http://esciencenews.com/articles/2008/07/14/sociological.research.shows .combined.impact.genetics.social.factors.delinquency.

40. Interview 41.

41. Interview 54.

42. His remarks also demonstrate social genomic researchers' desire for medical imperative to recede from large-scale research.

43. Guo and Tong, "Age at First Sexual Intercourse, Genes, and Social Context," 752.

44. Catherine Bliss, Field notes (SSGAC meeting, November 12–14, 2014).

45. Harrati, "Characterizing the Genetic Influences on Risk Aversion," 188.

46. Cornelius A. Rietveld, Sarah E. Medland, Jaime Derringer, Jian Yang, Tõnu Esko, Nicolas W. Martin, Harm-Jan Westra, et al., "GWAS of 126,559 Individuals Identifies Genetic Variants Associated with Educational Attainment," *Science* 340, no. 6139 (6–21) (2013): 1467.

47. Catherine Bliss, Field notes (SSGAC meeting, January 12–13, 2015).

48. Guang Guo, Xiao-Ming Ou, Michael Roettger, and Jean C. Shih, "The VNTR 2 Repeat in MAOA and Delinquent Behavior in Adolescence and Young Adulthood: Associations and MAOA Promoter Activity," *European Journal of Human Genetics* 16 (2008): 131.

49. Rietveld et al., "GWAS of 126,559 Individuals Identifies Genetic Variants Associated with Educational Attainment."

50. Also see Fowler and Dawes, "Two Genes Predict Voter Turnout." For prior pregenomic uses, see James H. Fowler, and Cindy D. Kam, "Beyond the Self: Social Identity, Altruism, and Political Participation," *Journal of Politics* 69, no. 3 (2007): 813–27.

51. C. Mitchell, "The Association of PTSD and Deployment with Telomere Erosion," University of Colorado, 2014, www.colorado.edu/ibs/cupc/conferences/IGSS_2014/abstracts/The%20Association%20of%20PTSD_Colter%20Mitchell.pdf.

52. Catherine Bliss, Field notes (IGSS meeting, October 9–10, 2014).

53. Thus far, social genomics researchers have not undertaken epigenetic studies, or studies of heritable transcriptional mutations that are not encoded in the genotype. Epigenetics currently relies on analysis of DNA methylation or histone modification, biochemical changes that are studied with mass spectrometry. Though the SSGAC is hoping to partner with epigeneticists at CHARGE, there are currently no models dealing with extragenomic factors.

54. Interview 33.

55. Interview 28.

56. Ibid. This researcher has been particularly instrumental in innovating lifecourse GWAS approaches, because "obesity, smoking and asthma . . . are the three primary drivers of poor health in the first part of the life course and of health disparities in children and young adults." Though his dual focus on early childhood and midlife aging has garnered him the accusation of "phenotypic promiscuity," he believes that social science will be left behind without a say in policy if it leaves GWAS studies of behavior to biomedicine.

57. Catherine Bliss, Field notes (IGSS meeting, October 9–10, 2014).

58. Eveline L. De Zeeuw, Catharina E. M. van Beijsterveldt, Tina J. Glasner, et al., "Polygenic Scores Associated with Educational Attainment in Adults Predict Educational Achievement and ADHD Symptoms in Children," *American Journal of Medical Genetics Part B* 165, no. 6 (2014): 517.

59. Interview 65.

60. Ibid.

61. Jonathan Beauchamp, David Cesarini, and Magnus Johannesson, "The Psychometric Properties of Measures of Economic Risk Preferences," *Journal of Economic Perspectives* 25, no. 4 (2011): 13.

62. Interview 53.

63. Interview 7.

64. Interview 5.

65. Morgan E. Levine, Eileen M. Crimmins, Carol A. Prescott, et al., "A Polygenic Risk Score Associated with Measures of Depressive Symptoms among Older Adults," *Biodemography and Social Biology* 60, no. 2 (2014): 200.

66. Interview 7.

67. Interview 11.

68. Matthijs J. H. M. Van der Loos et al., "The Molecular Genetic Architecture of Self-Employment," *PLoS ONE* 8, no. 4 (2013): e60542.

69. Eileen M. Crimmins, "Integrating Work from Genetics and the Social Sciences: Part Two," *Biodemography and Social Biology* 60 (2014).

70. Cornelius A. Rietveld, Dalton Conley, Nicholas Eriksson, et al., "Replicability and Robustness of Genome-Wide-Association Studies for Behavioral Traits," *Psychological Science* 25 (2014): 1984.

71. Ibid.

Chapter 4: Determining Difference

1. Bliss, *Race Decoded*.

2. See, for example, a race-based asthma study, the new NICHD Fetal Growth standards by race, and the racially slotted 1000 Genomes Project: A. M. Levin et al., "Nocturnal Asthma and the Importance of Race/Ethnicity and Genetic Ancestry," *American Journal of Respiratory Critical Care Medicine* 190, no. 3 (2014): 266–73; Louis Buck et al., "Racial/Ethnic Standards for Fetal Growth: The NICHD Fetal Growth Studies," *American Journal of Obstetrics and Gynecology* 213, no. 4 (2015): 449.e1–449.e41; Lizzie Wade, "What 2500 Sequenced Genomes Say about Humanity's Future," *Wired*, 2015.

3. Liz Harley, "Veritas Genetics Opens Chinese R&D Centre," *Frontline Genomics*, 2015, www.frontlinegenomics.com/1978/veritas-genetics-opens-chinese-rd-centre/; Jonathan Kahn, *Race in a Bottle* (New York: Columbia University Press, 2013); "Race-Based Medicine: A Recipe for Controversy," *Scientific American*, 2007, www.scientific american.com/article/race-based-medicine-a-recipe-for-controversy/; Matthew Herper, "Race-Based Medicine Arrives," *Forbes*, 2005, www.forbes.com/2005/05/10/cx_mh_0509 racemedicine.html.

4. Diana Fuss, in Gayatri Spivak, *Post Colonial Critic* (New York: Routledge, 1990).

5. Brown University, "'Warrior Gene' Predicts Aggressive Behavior after Provocation," *ScienceDaily*, 2009, www.sciencedaily.com/releases/2009/01/090121093343.htm.

6. Conservative Daily, "Science Daily: Whites Have Lowest Instance of MAOA-L Gene, Which Is Linked to Aggression, Violence, Crime, and Sexual Abuse," *Conservative Headlines*, 2010, http://conservative-headlines.com/2010/05/science-daily-whites -have-lowest-instance-of-maoa-l-gene-which-is-linked-to-aggression-violence-crime -and-sexual-abuse/.

7. Thomas B. Edsall, "Are Our Political Beliefs Encoded in Our DNA?," *New York Times, Opinionator* (blog), 2013, http://opinionator.blogs.nytimes.com/2013/10/01/are- our-political-beliefs-encoded-in-our-dna/.

8. Ibid.

9. Ibid.

10. Justin Glawe, "Identifying 'Criminal' Genes Will Never Prevent Violence but Might Help Explain It," *Vice*, 2014, www.vice.com/read/identifying-criminal-genes-will -never-prevent-violence-but-might-help-explain-it-1113.

11. Gautam Naik, "A Genetic Code for Genius?," *Wall Street Journal*, 2013, www.wsj.com/articles/SB10001424127887324162304578303992108696034.

12. Ibid.

13. Jason D. Boardman et al., "Ethnicity, Body Mass, and Genome-Wide Data," *Biodemography and Social Biology* (2010): 123–36.

14. Ibid.

15. Yi Li and Guang Guo, "Data Quality Control in Social Surveys Using Genetic Information," *Biodemography and Social Biology* (2014): 212–28.

16. Ibid.

17. Ibid.

18. Guang Guo et al., "Genetic Bio-ancestry and Social Construction of Racial Classification in Social Surveys in the Contemporary United States," *Demography* (2013): 141–72.

19. Ibid.

20. Ibid.

21. Jonathan Daw and Jason D. Boardman, "The Long Arm of Adolescence: School Health Behavioral Environments, Tobacco and Alcohol Co-use, and the 5HTTLPR Gene," *Biodemography and Social Biology* 60, no. 2 (2014): 117–36.

22. Daniel W. Belsky, Terrie E. Moffitt, and Avshalom Caspi, "Genetics in Population Health Science: Strategies and Opportunities," *American Journal of Public Health* 103, suppl. no. 1 (2013): S73–S83.

23. K. M. Beaver et al., "Evidence of a Gene X Environment Interaction in the Creation of Victimization: Results from a Longitudinal Sample of Adolescents," *International Journal of Offender Therapy and Comparative Criminology* (2007): 620–45.

24. Interviews 149, 152.

25. Beaver, "Evidence of a Gene X Environment Interaction in the Creation of Victimization." In another example from this team:

> Covariates. As mentioned above, a host of evidence suggests that the antisocial involvement of most individuals tends to peak in adolescence and decline shortly thereafter (i.e., AL offending patterns). In order to account for the age-graded nature of most antisocial behavior, we included age-coded continuously in years as a covariate in the current study. In addition, a consistent body of research has repeatedly demonstrated that African Americans are disproportionately represented in the criminal justice system, for a range of possible reasons. In order to avoid possible confounding due to race, then, we included this item as a dichotomous indicator, which is coded as 0 = non-African American and 1 = African American.

26. Benjamin W. Domingue, Jason Fletcher, Dalton Conley, and Jason D. Boardman, "Genetic and Educational Assortative Mating among US Adults," *Proceedings of the National Academy of Sciences* 111, no. 22 (2014): 7996–8000.

27. Ibid.

28. James H. Fowler, Jaime E. Settle, and Nicholas A. Christakis, "Correlated Genotypes in Friendship Networks," *Proceedings of the National Academy of Sciences* 108, no. 5 (2011): 1993–97.

29. Ibid.

30. Interview 23.

31. Ibid.

32. Interview 103.

33. Ibid.

34. Interview 2871.

35. Ibid.

36. McDermott et al., "Monoamine Oxidase A Gene (MAOA) Predicts Behavioral Aggression Following Provocation."

37. Wehby and von Hinke Kessler Scholder, "Genetic Instrumental Variable Studies of Effects of Prenatal Risk Factors."

38. Interview 216.

39. "Gangs, Guns … and Genes?," *Times Online*, 2009, www.thetimes.co.uk.

40. "Homo Administrans: Biology of Business," *Economist*, 2010, www.economist .com/node/17090697.

41. "Is There a Gene for Motherhood?," ABC News, *Good Morning America*, 2012, http://abcnews.go.com/blogs/lifestyle/2012/09/is-there-a-gene-for-motherhood/.

42. Ibid.

43. Florida State University, "'Warrior Gene' Reported Rife among Young Thugs," *World Science*, 2009, www.world-science.net/othernews/090605-maoa.htm.

44. Brown University, "'Warrior Gene' Predicts Aggressive Behavior after Provocation," *ScienceDaily*, 2009, www.sciencedaily.com/releases/2009/01/090121093343.htm.

45. Erik Voeten, "Are There Genes That Lead Men to Enjoy Fighting?," *Washington Post*, 2014, www.washingtonpost.com/news/monkey-cage/wp/2014/02/04/are-there -genes-that-lead-men-to-enjoy-fighting/.

46. Ibid.

47. George Will, "A Cure for Character," *Townhall*, 2010, http://townhall.com/ columnists/georgewill/2010/02/28/a_cure_for_character.

48. Brian Vastag, "'Callous-Unemotional' Children Often Grow Up to Lie, Fight, and Bully, Study Finds," *Washington Post*, 2011, www.washingtonpost.com/wp-dyn/content/ article/2011/02/21/AR2011022104004.html.

49. Gina Kolata, "Scientists to Seek Clues to Violence in Genome of Gunman in Newtown, Conn.," *New York Times*, 2012, www.nytimes.com/2012/12/25/science/scientists -to-seek-clues-to-violence-in-genome-of-gunman-in-newtown-conn.html.

50. Ibid.

51. Belinda L. Needham et al., "A Test of Biological and Behavioral Explanations for Gender Differences in Telomere Length: The Multi-ethnic Study of Atherosclerosis," *Biodemography and Social Biology* 60, no. 2 (2014): 156–73.

52. Bradshaw and Ellison, "Nature-Nurture Debate Is Over, and Both Sides Lost!"

53. Ibid.

54. Melissa A. Petkovsek, Brian B. Boutwell, Kevin M. Beaver, and J. C. Barnes, "Prenatal Smoking and Genetic Risk: Examining the Childhood Origins of Externalizing Behavioral Problems," *Social Science and Medicine* 111 (2014): 17–24.

55. Ibid.

56. James H. Fowler and Christopher T. Dawes, "In Defense of Genopolitics," *American Political Science Review* 107, no. 2 (2013): 362–74.

57. Daw and Guo, "Influence of Three Genes on Whether Adolescents Use Contraception, USA 1994–2002."

58. Ibid.

59. Muriel Zheng Fang, "Violating the Monotonicity Condition for Instrumental Variable-Dimorphic Patterns of Gene-Behavior Association," *Economics Letters* 122, no. 1 (2014): 59–63.

60. The article cites a number of animal model studies and then writes, "These dimorphic effects are hypothesized to be based on the biological foundation that sex hormones are similar to neurotrophins and regulate neurotransmitter plasticity."

61. Kevin M. Beaver, Christina Mancini, Matt DeLisi, and Michael G. Vaughn, "Resiliency to Victimization: The Role of Genetic Factors," *Journal of Interpersonal Violence* 26, no. 5 (2011): 874–98.

62. Yi Li, Hexuan Liu, and Guang Guo, "Does Marriage Moderate Genetic Effects on Delinquency and Violence?," *Journal of Marriage and Family* 77, no. 5 (2015): 1217–33.

63. Ibid.

64. Ibid.

65. Interview 88.

66. Interview 10.

67. Interview 95.

68. Interview 214.

69. Interview 2871.

70. Interview 48.

71. Interview 126.

72. "Study Suggests Genes Influence Men's Chances of Being Gay," *NY Daily News*, 2014, www.nydailynews.com/life-style/health/study-suggests-genes-influence-men-chances-gay-article-1.2013597.

73. Lindsay Tanner, "New Study Suggests Genetic Link for Male Homosexuality," *Huffington Post*, 2014, www.huffingtonpost.com/2014/11/17/genetic-link-male-homosexuality_n_6171244.html.

74. Simon Copl, "Born This Way? Society, Sexuality and the Search for the 'Gay Gene.'" *Guardian*, 2015, www.theguardian.com/science/blog/2015/jul/10/born-this-way-society-sexuality-gay-gene.

75. "Gay Gene Discovery Has Good and Bad Implications," *New Scientist, New Science Opinion Blog*, 2014, www.newscientist.com/article/mg22429963-700-gay-gene-discovery-has-good-and-bad-implications.

76. Sarah Knapton, "Being Homosexual Is Only Partly due to Gay Gene, Research Finds," *Telegraph*, 2014, www.telegraph.co.uk/news/science/science-news/10637532/Being-homosexual-is-only-partly-due-to-gay-gene-research-finds.html.

77. "Sexual Offending Runs in Families," Today, BBC Radio 4, 2014, www.bbc.co.uk/programmes/po2nrol7.

78. Ibid.

79. Conor Gaffey, "Relatives of Rapists More Likely to Offend, Finds Study," *Newsweek*, 2015, http://europe.newsweek.com/genetics-play-decisive-role-sexual-offending-321160.

80. Sarah Knapton, "Sex Offending Is Written in DNA of Some Men, Oxford University Finds," *Telegraph*, 2015, www.telegraph.co.uk/news/science/science-news/11522366/Sex-offending-is-written-in-DNA-of-some-men-Oxford-University-finds.html.

81. "Study Finds Genetic Component to Sexual Offending," ABC 7, 2015, www.abc-7.com/story/28771097/genetics-can-make-people-more-likely-to-commit-sex-offenses-study-says.

82. Don Melvin, "Study: Genetics Can Increase Likelihood of Sex Offense," CNN, 2015, www.cnn.com/2015/04/09/europe/sex-offenders-genetic-predisposition/index.html.

83. "Study into Gene Link in Sex Attackers," BBC News, 2015, www.bbc.com/news/uk-32221589.

84. "Genes 'Play Role in Sex Attackers,'" *Belfast Telegraph*, 2015, www.belfasttelegraph.co.uk/news/uk/genes-play-role-in-sex-attackers-31128616.html.

85. Emily Underwood, "Reality Check: Is Sex Crime Genetic?," *Science*, 2015, http://news.sciencemag.org/brain-behavior/2015/04/reality-check-sex-crime-genetic.

86. Richard A. Friedman, "Infidelity Lurks in Your Genes," *New York Times*, 2015, www.nytimes.com/2015/05/24/opinion/sunday/infidelity-lurks-in-your-genes.html.

87. Tony Bravo, "Are Your Genes the Reason You Are Single?," *SF Gate*, 2014, http://blog.sfgate.com/relationships/2014/11/21/are-your-genes-the-reason-youre-single/?utm_source=http://blog.sfgate.com/topdown/2014/04/?utm_source=http://blog.sfgate.com/topdown/page/4/?auth=380&o=179&utm_medium=referral&utm_campaign=offpage-widget-dealer-spotlight&utm_content=http://blog.sfgate.com/topdown/2014/04/&utm_medium=referral&utm_campaign=offpage-widget-dealer-spotlight&utm_content=http://blog.sfgate.com/relationships/2014/11/21/are-your-genes-the-reason-youre-single/.

88. "Video: Instant Chemistry: Putting Relationships to a Genetic Test," ABC News, *Nightline*, 2014, http://abcnews.go.com/Nightline/video/instant-chemistry-putting-relationships-genetic-test-24462358.

89. Ibid.

90. Ruthie Friedl, "Is Genetic Testing the Future of Online Dating?," *Elle*, 2014, www.elle.com/news/lifestyle/genetic-testing-online-dating.

91. Ellen McCarthy, "'Being Away from Each Other Was Kind of a Blessing,'" *Washington Post*, 2010, www.washingtonpost.com/wp-dyn/content/article/2010/01/21/AR2010012105182.html.

92. Ibid. See also Erik Sherman, "Computer Dating Meets DNA Analysis," CBS News, *MoneyWatch*, 2014, www.cbsnews.com/news/computer-dating-meets-dna-analysis/.

93. "Genetic Love Match? Dating Sites Try DNA Tests," NBC News, 2009, www.nbcnews.com/id/33893470/ns/health-sexual_health/t/genetic-love-match-dating-sites-try-dna-tests/.

94. Guo and Tong, "Age at First Sexual Intercourse, Genes, and Social Context."

95. Brian B. Boutwell, J. C. Barnes, and Kevin M. Beaver, "Life-Course Persistent Offenders and the Propensity to Commit Sexual Assault," *Sexual Abuse: A Journal of Research and Treatment* 25, no. 1 (2013): 69–81.

96. Guo and Tong, "Age at First Sexual Intercourse, Genes, and Social Context."

97. Boutwell et al., "Life-Course Persistent Offenders and the Propensity to Commit Sexual Assault."

98. "Thematic Session: Networks and Sexual Behaviors," ASA, 2015, www.asanet.org/AM2015/Documents/Friday%20August%2021%20and%20Saturday%20August%2022.pdf.

99. Guo et al., "Genetic Bio-ancestry and Social Construction of Racial Classification in Social Surveys in the Contemporary United States."

100. Conley, "Learning to Love Animal (Models) (or) How (Not) to Study Genes as a Social Scientist."

101. Ibid.

102. Peter K. Hatemi and Rose McDermott, "The Genetics of Politics: Discovery, Challenges, and Progress," *Trends in Genetics* 28, no. 10 (2012): 525–33.

103. Ibid.

104. Interview 99.

105. Ibid.

106. Ibid.

107. Interview 731.

108. Jayaratne et al., "White Americans' Genetic Lay Theories of Race Differences and Sexual Orientation."

109. Morning, *Nature of Race.*

110. Ibid.

111. See, for example, Reardon, *Race to the Finish*; Richardson, *Sex Itself*; Roberts, *Fatal Invention.*

112. Nelson, *Social Life of DNA,* 3.

113. Brodwin, "Genetics, Identity, and the Anthropology of Essentialism."

Chapter 5: The Breakthrough

1. Interview 1.

2. Interview 1.

3. Interview 91.

4. Interview 43.

5. Interview 4.

6. Interview 124.

7. Interview 91.

8. Interview 9.

9. Interview 115.

10. Interview 8.

11. Interview 30.

12. Interview 10.
13. Interview 27.
14. Interview 25.
15. Interview 28.
16. Interview 28.
17. Interview 116.
18. Interview 31.
19. Interview 93.
20. Interview 42.
21. Interview 50.
22. Interview 44.
23. Interview 85.
24. Interview 27.
25. Interview 85.
26. Interview 111.
27. Interview 45.
28. Interview 1.
29. Interview 9.
30. Interview 8.
31. Interview 104.
32. Interview 3.
33. Interview 1.
34. Interview 109.
35. Interview 8.
36. Interview 121.
37. Interview 8.
38. Interview 92.
39. Interview 90.
40. Panofsky, *Misbehaving Science*.
41. Interview 27.
42. Interview 126.
43. Interview 89.
44. Interview 126. This researcher saw social genomics as a protector against bad bioscience. She continued:

> It's like you do understand the big pharmaceutical is doing it already and they're selling it to you and you're buying into it without any objection whatsoever. The only balance against those companies, who by the way are only doing it for money—don't believe that they're doing it for your health—is me. And you think I'm the one who's evil in this story.

45. Interview 43.
46. Interview 43.
47. Interview 29.
48. Interview 30.

49. Interview 40.
50. Interview 87.
51. Ibid.
52. Interview 59.
53. Interview 116.
54. Interview 120.
55. Interview 3.
56. Interview 1.
57. Interview 40.
58. Interview 60.
59. Interview 109.
60. Interview 97. He went on:

So, I mean, you encounter that and I think the most frustrating thing from my perspective is not encountering opposition because that's good science. I mean this rampant failure to try and engage with what we're actually saying. Instead there's a lot of straw man stuff. So the notion that anybody who's doing this is engaged in biological determinism gets me frustrated because it just shows a failure to have read this stuff before you spoke up. If I say, "Ideology is forty percent heritable," and you say, "You're a biological determinist!" I think, "In what world is forty percent deterministic?" Political science, more so than some others, has been environmentally deterministic. And that doesn't seem to bother them.

Indeed, frustration with "environmental determinism" often came up in a majority of my interviews and conversations with social genomics researchers.

61. Interview 1.
62. Interview 45.
63. Interview 115.
64. Interview 121.
65. Interview 89.
66. Interview 11.
67. Interview 45.
68. Interview 108.
69. Interview 112
70. Interview 117.
71. Interview 115.
72. Interview 126.
73. Interview 118.
74. Interview 46.
75. Interview 50.
76. Interview 91.
77. Interview 6.
78. Interview 3.
79. Interview 127.

80. Interview 128.

81. For example, one student at an elite midwestern research institute sought out partners at a sister Research-1 school before finding out that the work he had been doing was a replica of something another scientist had done before (Interview 104).

82. Interview 97.

83. Interview 26.

84. Interview 111.

Chapter 6: A Bigger, Better Science

1. Catherine Bliss, Field notes (IGSS meeting, 2013).

2. "Integrating Genetics and the Social Sciences," *Biodemography and Social Biology* 57, no. 1 (2011).

3. "Society, Genetics, and Health," *American Journal of Public Health* 103 (October 2013): S1, http://ajph.aphapublications.org/action/showLargeCover?issue=40100437.

4. "Policy Analysis and Genetics," *Journal of Policy and Management* 34, no. 3 (2015).

5. Rietveld, "GWAS of 126,559 Individuals Identifies Genetic Variants Associated with Educational Attainment."

6. Aysu Okbay et al., "Genome-wide Association Study Identifies 74 Loci Associated with Educational Attainment," *Nature* 533 (2016): 539–42.

7. See Fowler and Dawes, "Two Genes Predict Voter Turnout"; James H. Fowler and Cindy D. Kam, "Beyond the Self: Social Identity, Altruism, and Political Participation," *Journal of Politics* 69 (2008): 813–27; Rose McDermott, James H. Fowler, and Oleg Smirnov, "On the Evolutionary Origin of Prospect Theory Preferences," *Journal of Politics* 70 (2008): 335–50; J. H. Fowler and D. Schreiber, "Biology, Politics, and the Emerging Science of Human Nature," *Science* 332 (2008): 912–14.

8. Catherine Bliss, Field notes (SSGAC meeting, 2013).

9. Interview 7.

10. Interview 102.

11. Interview 88.

12. Interview 2.

13. Interview 48.

14. Interview 108.

15. Interview 43.

16. Ibid.

17. Interview 112.

18. Interview 87.

19. Interview 112.

20. Interview 114.

21. Interview 116.

22. Interview 216.

23. One editor of a leading sociological journal said that she was committed to raising genetics up in the estimation of her field (Interview 40). Press releases have been an important part of the publishing process. One principal investigator who has gotten

articles published in a number of specialty medical journals but hasn't gotten his team's publications into *Science, Nature,* or *PNAS* said that with the exception of one publication in a specialty medical journal with a high impact factor, "all the genetic genetics paper I've written have had an editorial to go with them" (Interview 25). For the exception, his home institution and the journal issued dual press releases, which sparked media coverage, and eventually led to him being on TV.

24. "Social and Economic Sciences," National Science Foundation, 2015, www.nsf .gov/div/index.jsp?div=SES.

25. Ibid.

26. John Alford, John Hibbing, and Kevin Smith, "Genes and Politics: Providing the Necessary Data," *Grantome,* 2007, http://grantome.com/grant/nsf/ses-0721379; John Alford, John Hibbing, and Kevin Smith, "Investigating the Genetic Basis of Economic Behavior," *Grantome,* 2007, http://grantome.com/grant/nsf/ses-0721776; John Alford, "DHB: Identifying Biological Influences on Political Temperaments," NSF Awards, 2009, www.nsf.gov/awardsearch/showaward?awd_id=0826867.

27. Quamrul Ashraf, "Genetic Diversity and the Wealth of Nations," NSF Awards, 2013, www.nsf.gov/awardsearch/showaward?awd_id=1338426.

28. Chris Dawes and James Fowler, "A Genome-Wide Association Study of Voter Turnout," NSF Awards, 2010, www.nsf.gov/awardsearch/showAward?AWD_ID=1024064.

29. James Fowler, "The Genetic Basis of Social Networks and Civic Engagement," NSF Awards, 2008, www.nsf.gov/awardsearch/showAward?AWD_ID=0719404.

30. Jaime Settles, "Understanding the Mechanisms for Disengagement Contentious Political Interaction," NSF Awards, 2014, www.nsf.gov/awardsearch/showAward?AWD _ID=1423788.

31. "Research Supported by the NICHD," NIH, 2015, www.nichd.nih.gov/research/ by-nichd/Pages/index.aspx.

32. "Program Project/Center Grants," NIH, 2015, http://grants.nih.gov/grants/fund ing/funding_program.htm#PSeries.

33. Jason Boardman, "The Social Determinants of Genetic Expression: A Life-Course Perspective," *Grantome,* 2009, http://grantome.com/grant/NIH/K01-HD050336-02.

34. Matthew Mcqueen and Jason Boardman, "Social Demographic Moderation of Genome Wide Associations for Body Mass Index," *Grantome,* 2010, http://grantome .com/grant/NIH/R01-HD060726-01A2. This grant proposes pharmaceutical and diagnostic development.

35. Jason Boardman and Benjamin Domingue, "The Social and Genetic Epidemiology of Health Behaviors: An Integrated Approach," *Grantome,* 2013, http://grantome. com/grant/NIH/R21-HD078031-02.

36. Jason Fletcher, "Examining the Sources and Implications of Genetic Homophily in Social Networks," *Grantome,* 2012, http://grantome.com/grant/NIH/R21-HD 071884-03.

37. Nicholas Christakis, "Networks and Neighborhoods," *Grantome,* 2009, http:// grantome.com/grant/NIH/P01-AG031093-01.

38. Jeremy Freese, Robert Hauser, and Pamela Herd, "A Longitudinal Resource for Genetic Research in Behavioral and Health Sciences," *Grantome*, 2015, http://grantome.com/grant/NIH/R01-AG041868-03.

39. Michael Shanahan, "Genetic Risk, Pathways to Adulthood, and Health Inequalities," *Grantome*, 2010, http://grantome.com/grant/NIH/R01-HD061622-01A1.

40. "Division of Behavioral and Social Research," NIH, 2015, www.nia.nih.gov/research/dbsr.

41. "Consolidator Grants," European Research Council, 2012, https://erc.europa.eu/consolidator-grants.

42. "Unravelling the Genetic Influences of Reproductive Behaviour and Gene-Environment Interaction," European Research Council, 2014, https://erc.europa.eu/unravelling-genetic-influences-reproductive-behaviour-and-gene-environment-interaction.

43. Philipp Koellinger, "ERC Consolidator Grant for Philipp Koellinger," Amsterdam Business School, University of Amsterdam, 2015, http://abs.uva.nl/news-events/content/2015/03/erc-consolidator-grant-for-philipp-koellinger.html.

44. "Research and Studies for the Office of Net Assessment," Federal Grants, 2016, www.federalgrants.com/Research-and-Studies-for-the-Office-of-Net-Assessment-OSD-NA-17674.html.

45. Kevin Beaver, "Intersection of Genes, the Environment, and Crime and Delinquency: A Longitudinal Study of Offending," National Criminal Justice Reference Service, 2010, www.ncjrs.gov/app/publications/abstract.aspx?ID=253671.

46. "DoD Announces Appointment of James Baker as Director of the Office of Net Assessment," US Department of Defense, 2015, www.defense.gov/News/News-Releases/News-Release-View/Article/605502/dod-announces-appointment-of-james-baker-as-director-of-the-office-of-net-asses. See also Rose McDermott, "Curriculum Vitae," Watson Institute, Brown University, 2015, http://watson.brown.edu/people/faculty/mcdermott.

47. "Army Study to Assess Risk and Resilience in Servicemembers (Army STARRS): A Partnership between NIMH and the U.S. Army," NIH, 2014, www.nimh.nih.gov/health/topics/suicide-prevention/suicide-prevention-studies/army-study-to-assess-risk-and-resilience-in-servicemembers-army-starrs-a-partnership-between-nimh-and-the-us-army.shtml; "Army Study to Assess Risk and Resilience in Servicemembers (Army STARRS)," Military Health Matters, 2015, www.militaryhealthmatters.com/milhealths-research/army-study-to-assess-risk-and-resilience-in-servicemembers-army-starrs.

48. "Funding Opportunities," Russell Sage Foundation, 2015, www.russellsage.org/research/categories/requests-proposals.

49. "Funding Opportunity: Program on Behavioral Economics," Russell Sage Foundation, 2015, www.russellsage.org/research/funding/behavioral-economics.

50. "GxE and Health Inequality over the Life Course," Russell Sage Foundation, 2015, www.russellsage.org/awarded-project/gxe-and-health-inequality-over-life-course.

51. "Role of Genetics in Preference Formation," Russell Sage Foundation, 2006, www.russellsage.org/awarded-project/role-genetics-preference-formation.

52. "Biosocial Pathways of Well Being across the Course," Russell Sage Foundation, 2015, www.russellsage.org/publications/category/current_rfa_rsfjournal/biosocial-pathways.

53. "Summer Institute in Social Science Genomics," Russell Sage Foundation, 2016, www.russellsage.org/research/funding/social-science-genomics.

54. "John Simon Guggenheim Foundation, About the Fellowship," Guggenheim Foundation, 2015, www.gf.org/about/fellowship/.

55. "John Simon Guggenheim Foundation, Dalton Conley," Guggenheim Foundation, 2011, www.gf.org/fellows/all-fellows/dalton-conley/.

56. Jason Fletcher, "Interconnected Contexts: The Interplay between Genetics and Social Settings in Youth Development," William T. Grant Foundation, 2012, http://wtgrantfoundation.org/browsegrants#/grant/184625.

57. Jason Boardman, "Genetic Aspects of Psychological Resiliency among US Adults," University of Colorado, Boulder, 2006, https://behavioralscience.colorado.edu/grant/genetic-aspects-of-psychological-resiliency-among-u-s-adults-midus.

58. Colter Mitchell, "Applying Whole Genome Data to Common Social Science Issues," Michigan Center on Demography of the Aging, 2014, http://micda.psc.isr.umich.edu/project/detail/36316.

59. Jason Fletcher, "Exploring the Relevance of Gene-Environment Interactions for the Social Sciences," Columbia University, 2011.

60. "About NBER," National Bureau of Economic Research, 2015, www.nber.org/info.html.

61. Interview 239.

62. Interview 216.

63. Interview 239.

64. Interview 632.

65. Ibid.

66. Interview 240.

67. Interview 216.

68. Interview 239.

69. Ibid.

70. Interview 216.

71. Ibid.

72. Interview 239.

73. Interview 240.

74. Interview 648.

75. Ibid.

76. *Interview 241.*

77. Ibid.

78. Interview 119.

79. Interview 49.

80. Ibid.

81. Interview 732.

82. Interview 733.

83. Interview 642.

84. Ibid.

85. Ibid.

86. Ibid.

87. Interview 1040.

88. Jade D'Alpoim Guedes et al., "Is Poverty in Our Genes?: A Critique of Ashraf and Galor, 'The "Out of Africa" Hypothesis, Human Genetic Diversity, and Comparative Economic Development,' *American Economic Review* (Forthcoming) 1," *Current Anthropology* 54, no. 1 (2013): 71–79.

89. Ibid.

90. Ewan Callaway, "Economics and Genetics Meet in Uneasy Union," *Nature*, 2012, www.nature.com/news/economics-and-genetics-meet-in-uneasy-union-1.11565.

91. Ibid.

92. Interview 734.

93. Ibid.

94. Ibid.

95. Ibid.

96. Ibid.

97. Interview 773.

98. I must also mention that this expert voiced the additional fear that the authors of the 2010 critique held that racist interpretations and implications were in our future: "I fear that we are not only going to get really naive social science but we are going to get really naive human population variation genetics, that is tracking types of social behaviors and then signifying in racial, national, ethnic, continental kinds of narrative, not to mention sex differences."

99. Interview 649.

100. Ibid.

101. Interview 771.

102. Ibid.

103. Ibid.

104. Interview 772.

105. http://convention2.allacademic.com/one/asc/asc15/index.php?cmd=Online+Program+View+Event&selected_box_id=197015&PHPSESSID=gi7p7itcscpcgdf9uqhcqs05j5; http://paa2015.princeton.edu/sessions/21; http://editorialexpress.com/conference/RES2015/program/RES2015.html#131.

106. Jonathan Pierre Beauchamp, "Essays in Economics, Genetics, and Psychology," Harvard University, ProQuest Dissertations Publishing, 2011, http://search.proquest.com/docview/877616800/abstract?accountid=14525.

107. Hexuan Liu, "Integration of Sociology with Genomics in Studies of Delinquency and Violence, and Social Stratification and Mobility," Department of Sociology,

Graduate Students on the Market, University of North Carolina, 2015, http://sociology
.unc.edu/about-us/people/graduate-students-on-the-job-market/; see also www.colorado
.edu/ibs/cupc/conferences/IGSS_2015/papers.html.

108. Tinbergen Institute, "Genoeconomics PhD Course," SSGAC, 2014, http://ssgac
.org/documents/GenoeconomicsPhdcourseTI2014-2015.pdf.

109. "Courses," Department of Sociology, University of North Carolina at Chapel
Hill, 2015, http://sociology.sites.unc.edu/graduate-program/courses/.

110. Nicola Barben, "Sociology and Genetics," University of Oxford, 2014, www
.sociology.ox.ac.uk/course/sociology-and-genetics.html.

111. Sheila Jasanoff, *States of Knowledge: The Co-production of Science and the Social
Order* (London: Taylor & Francis, 1995).

Chapter 7: Applied Science

1. "The History of Genetic Fingerprinting," Department of Genetics, University of
Leicester, 2015, www2.le.ac.uk/departments/genetics/jeffreys/history-gf.

2. Ibid.

3. John Eligon and Timothy Williams, "Police Program Aims to Pinpoint Those Most
Likely to Commit Crimes," *New York Times*, 2015, www.nytimes.com/2015/09/25/us/
police-program-aims-to-pinpoint-those-most-likely-to-commit-crimes.html?smprod
=nytcore-iphone&smid=nytcore-iphone-share&_r=0.

4. Guatam Naik, "A Genetic Code for Genius?," *Wall Street Journal*, 2013, www.wsj
.com/articles/SB10001424127887324162304578303992108696034.

5. Andrew Levy, "Success DOES Depend on Your Parents' Intelligence: Exam Results Are Influenced by Genes, Not Teaching," *Daily Mail Online*, 2013, www.dailymail
.co.uk/sciencetech/article-2377575/Success-DOES-depend-parents-intelligence-GCSE
-results-influenced-genes-teaching.html.

6. Alexandra Ossola, "Is Being Good at Science a Genetic Trait?," *Atlantic*, 2014,
www.theatlantic.com/education/archive/2014/11/is-being-good-at-science-a-genetic
-trait/382287/.

7. Ibid.

8. Ibid.

9. "Genes May Help Identify Children with Future Psychological Problems," Fox
News, 2015, www.foxnews.com/health/2015/01/07/genes-may-help-identify-children-
with-future-psychological-problems.html.

10. Adrian Raine, "Opinion: Unlocking Crime Using Biological Keys," CNN, 2013,
www.cnn.com/2013/05/03/health/biology-crime-violence/index.html.

11. Jennifer Kahn, "Trouble at Age 9," *New York Times*, 2012, www.nytimes.com/
2012/05/13/magazine/trouble-at-age-9.html.

12. Maggie Fox, "Study Finds Genetic Link to Violence, Delinquency," Reuters,
2008, www.reuters.com/article/2008/07/14/us-delinquents-genes-idUSN144487242008
0714#mkKOLGJizJykWzuS.97.

13. Kolata, "Scientists to Seek Clues to Violence in Genome of Gunman in Newtown, Conn."

14. "Genes Linked to Violent Behaviour," *ABC Science*, 2014, www.abc.net.au/science/articles/2014/10/29/4116967.htm.

15. Sarah Knapton, "Violence Genes May Be Responsible for One in 10 Serious Crimes," *Telegraph*, 2014, www.telegraph.co.uk/news/science/science-news/11192643/Violence-genes-may-be-responsible-for-one-in-10-serious-crimes.html.

16. "Study Finds Genetic Component to Sexual Offending," *ABC 7 News*, 2015, www.abc-7.com/story/28771097/genetics-can-make-people-more-likely-to-commit-sex-offenses-study-says.

17. Conor Gaffey, "Relatives of Rapists More Likely to Offend, Finds Study," *Newsweek*, 2015, http://europe.newsweek.com/genetics-play-decisive-role-sexual-offending-321160.

18. Rhoda Lee, "Experts Identify Genetic Biomarkers Linked to PTSD," *Tech Times*, 2015, www.techtimes.com/articles/38950/20150312/experts-identify-genetic-biomarkers-linked-to-ptsd.htm.

19. Caelainn Hogan, "Blood Test Could Predict Risk of Suicide," *Washington Post*, July 30, 2014, www.washingtonpost.com/news/to-your-health/wp/2014/07/30/blood-test-could-predict-risk-of-suicide/.

20. "Strangers Can Spot Genetic Disposition for Empathy," NPR, *Talk of the Nation*, 2011, www.npr.org/2011/11/18/142512092/strangers-can-spot-genetic-disposition-for-empathy.

21. Nicole Oran, "Next Time You Go to the Doctor—Open Your Mouth, Say 'Ahh' and Take a Personality Test," *Medcity News*, 2014, http://medcitynews.com/2014/12/next-time-go-doctor-open-mouth-say-ahh-take-personality-test/.

22. Nathaniel McGregor, "Seven New Genes Linked to Anxiety Disorders," *The Conversation*, 2015, http://theconversation.com/seven-new-genes-linked-to-anxiety-disorders-42835.

23. Randy O. Frost and Gail Steketee, "Hoarding: When Too Much 'Stuff' Causes Grief," NPR, 2010, www.npr.org/templates/story/story.php?storyId=126386317.

24. Tara Parker-Pope, "The Fat Trap," *New York Times*, 2011, www.nytimes.com/2012/01/01/magazine/tara-parker-pope-fat-trap.html.

25. Sarah Knapton, "Being Homosexual Is Only Partly due to Gay Gene, Research Finds," *Telegraph*, 2014, www.telegraph.co.uk/news/science/science-news/10637532/Being-homosexual-is-only-partly-due-to-gay-gene-research-finds.html.

26. Erik Sherman, "Computer Dating Meets DNA Analysis," *CBS News*, 2014, www.cbsnews.com/news/computer-dating-meets-dna-analysis/.

27. Debra Wilson, *Genetics, Crime and Justice* (Cheltenham, UK: Edward Elgar, 2015).

28. Michelle Henery, "Killer Blamed His Family History," *Times* (London), 2002, www.lexisnexis.com.ucsf.idm.oclc.org/lnacui2api/api/version1/getDocCui?lni=46WT-VMR0-00GN-Y051&csi=8411&hl=t&hv=t&hnsd=f&hns=t&hgn=t&oc=00240&perma=true.

29. Han G. Brunner, "Monoamine Oxidase and Behaviour," *Annals of Medicine* 27, no. 4 (1995): 431–32.

30. The Nuffield Council on Bioethics encouraged the use of genetic information. The council said that when there is "enough" genetic information it should absolutely

be used to mitigate sentences. Although it may not affect charges it could change sentencing, linking more offences to mental and genetic predisposition. "While a genetic predisposition to antisocial behaviour should not be a defence, such information could 'assist in determining degrees of blame,' according to a report by a panel of scientists, philosophers, ethicists and lawyers"; Mark Henderson, "Criminal Gene 'Should Mean Lighter Sentence,'" *Times*, 2002.

31. Barbara Bradley Hagerty, "Can Your Genes Make You Murder?," NPR, 2010, www.npr.org/templates/story/story.php?storyId=128043329.

32. Emiliano Feresin, "Lighter Sentence for Murderer with 'Bad Genes,'" *Nature News*, 2009, www.nature.com/news/2009/091030/full/news.2009.1050.html.

33. Peter McKnight, "Genetics Alone Cannot Explain Behavior," *Vancouver Sun*, 2013, www.vancouversun.com/health/Genetics+alone+cannot+explain+behaviour/7874720/story.html.

34. Andrew Solomon, "The Reckoning," *New Yorker*, 2014, www.newyorker.com/magazine/2014/03/17/the-reckoning.

35. Elijah Wolfson, "My Genes Did It!," *Newsweek*, 2014, www.newsweek.com/2014/03/14/my-genes-did-it-247951.html.

36. People v. Adams, No. S118045 (2014).

37. Tanzi v. Secretary, Florida Department of Corrections, No. 13-12421 (2014).

38. Barrett v. United States of America, No. 12-7806 (2015).

39. Chatman v. Walker, S15A0260 (2015).

40. Delgado v. State of Florida, No. SC12-579 (2015).

41. Emily Chang, "In China, DNA Tests on Kids ID Genetic Gifts, Careers," CNN, 2009, http://edition.cnn.com/2009/WORLD/asiapcf/08/03/china.dna.children.ability/.

42. Interview 1020.

43. Ibid.

44. Interview 1004.

45. Ibid.

46. Ibid.

47. Interview 1043.

48. Ibid.

49. Interview 1003.

50. Ibid.

51. Interview 1019.

52. Ibid.

53. Ibid.

54. Interview 1042.

55. Ibid.

56. Interview 1045

57. Ibid.

58. Interview 1044.

59. Cornelius A. Rietveld et al., "Common Genetic Variants Associated with Cognitive Performance Identified Using the Proxy-Phenotype Method," *Proceedings of the National Academy of Sciences* 111, no. 38 (2014): 13790–94.

60. For example, consider a program that aims at improving the education of disadvantaged children. Such a program could easily be very expensive, so a government may decide to run a small, randomized field experiment to test the effectiveness of the intervention before implementing it on a larger scale. Research that aims to evaluate the effectiveness of this intervention may benefit from including variables derived from genetic data (e.g., polygenic risk scores).

61. Catherine Bliss, Field notes (IGSS meeting, 2014).

62. Catherine Bliss, Field notes (ASA meeting, 2015).

63. Dalton Conley, "What If Tinder Showed Your IQ?," *Nautilus*, 2015, http://nautil.us/issue/28/2050/what-if-tinder-showed-your-iq.

64. Interview 3.

65. Interview 107.

66. Interview 747.

67. Interview 5.

68. Interview 94.

69. Ibid.

70. Interview 651.

71. Interview 741.

72. Interview 741.

73. Interview 2.

74. Interview 749.

75. Ibid.

76. Ibid.

77. Ibid.

78. Interview 654.

79. Interview 742.

80. Interview 746.

81. Interview 48.

82. Interview 743.

83. Interview 31.

84. Interview 652.

85. Interview 653.

86. Ibid.

87. Interview 744.

88. Ibid.

89. Interview 99.

90. Levy, "Success DOES Depend on Your Parents' Intelligence."

91. Ajai Raj, "Humans Genetically Engineered to Be Super Intelligent Could Have an IQ of 1000," *GateHouse Media*, 2014.

92. Interview 650.

93. Interview 215

94. Interview 124.

95. Interview 96.

96. Interview 88.

97. Ibid.

98. Interview 107.

99. Interview 123.

100. Mildred K. Cho and Pamela Sankar, "Forensic Genetics and Ethical, Legal, and Social Implications beyond the Clinic," *Nature Genetics 36 (2004): S8–S12.*

101. Paul Brodwin, "Genetics, Identity, and the Anthropology of Essentialism," *Anthropological Quarterly* 75, no. 2 (2002): 232–330.

102. Robert Merton, "The Self-Fulfilling Prophesy," *Antioch Review* 8, no. 2 (1945): 193–210; "The Unanticipated Consequences of Social Action," *American Sociological Review* 1, no. 6 (1936): 894–904.

Chapter 8: The Business of Sociogenomics

1. "Children's DNA Discovery," Smart DNA Testing, 2015, www.dnatests.me/parents-children.php.

2. Ibid.

3. "My Child's DNA Insights Kit," Makings of Me: Know Your DNA, 2013, www.themakingsofme.com/themakingsofme-test-kit-for-kids.html.

4. "Innate Talent and Behavioral Genetic Analysis," CNN, 2013, www.youtube.com/watch?v=jsRFr37lGZg; old link: http://dnamaptech.com/innate-talent-behavioural-genetic-analysis/.

5. Troy Duster, *Backdoor to Eugenics* (New York: Routledge, 2003).

6. "Inborn Talent Genetic Test," Map My Gene, 2009, www.mapmygene.com/inborn.htm.

7. Ibid.

8. Hernstein and Murray, *Bell Curve.*

9. James Watson, "To Question Genetic Intelligence Is Not Racism," Map My Gene, 2007, www.mapmygene.com/to-question-genetic-intelligence-is-not-racism.html.

10. "Learning from Mistakes," GenePlanet Personal Genetics, 2012, www.geneplanet.com/genetic-analysis/list-of-analyses/learning-from-mistakes.html#.

11. "Episodic Memory," GenePlanet Personal Genetics, 2012, www.geneplanet.com/genetic-analysis/list-of-analyses/episodic-memory.html.

12. H. Scott, "Crowdsourcing Health Conditions," *23andMe Blog,* 2012, http://blog.23andme.com/health-traits/crowdsourcing-health-conditions/.

13. Community, *23andMe Blog,* 2010–15, http://blog.23andme.com.

14. Ibid.

15. Community, "Specific Genes Linked to Big Brains and Intelligence," *23andMe Blog,* 2012, http://blog.23andme.com.

16. Community, "Is There Anyone Else with 2 Copies of Gene for Intelligence?," *23andMe Blog,* 2012, http://blog.23andme.com.

17. Community, "Scientists at Kings College, London Have Found the First Gene Which Appears to Be Linked to Intelligence," *23andMe Blog,* 2014, http://blog.23andme.com.

18. Community, "Specific Genes Linked to Big Brains and Intelligence," *23andMe Blog,* 2012, http://blog.23andme.com.

19. "Are You a Warrior?," Family Tree DNA, 2001–15, www.familytreedna.com/landing/warrior-gene.aspx.

20. Ibid.

21. C. Burdick, "Finding Warrior Roots for Martial Arts," *UConn Today*, 2010, http://today.uconn.edu/2010/01/finding-warrior-roots-for-martial-arts/.

22. "Lifestyle Gene Test Package," *Gentest*, 2015, www.gene-tests.org/factoid.

23. Ibid.

24. "Do You Have the Warrior Gene?," iGENEA, 2010, www.igenea.com/en/warrior-gene.

25. "Information That Women and Men Can Discover from Their Mitochondrial DNA," iGENEA, 2015, www.igenea.com/en/mitochondrial-dna.

26. "Products: Autosomal Transfers," Family Tree DNA, 2001–15, www.familytreedna.com/products.aspx.

27. "Pathway Fit," Pathway Genomics, 2015, www.pathway.com/pathway-fit-promo/.

28. "Pathway Genomics Launches $99 Pathway Fit Genetic Test for Limited Time," *Business Wire*, 2015, www.businesswire.com/news/home/20150806005379/en/Pathway-Genomics-Launches-99-Pathway-Fit-Genetic. See also "Pathway Fit," Pathway Genomics, 2015, www.pathway.com/pathway-fit-promo/.

29. "Test Specifications," Phenom Bioscience, 2015, www.phenombio.com/#mygpro-dna-test-details.

30. "Pathway Fit," Pathway Genomics, 2015, www.pathway.com/pathway-fit-promo/.

31. "You May Not Be Obese Because of Genes or Habits—It May Just Be Time," Science 2.0, *Scientific Blogging*, 2014, www.science20.com/news_articles/you_may_not_be_obese_because_of_genes_or_habits.

32. "Health Woman DNA Insight," Pathway Genomics, 2015, www.pathway.com/healthy-woman-dna-insight/.

33. Ibid.

34. "Science," Simplified Genetics, 2015, http://simplifiedgenetics.com/science.html.

35. "Genetic Test of Athletic Abilities," Sports Gene, 2014, www.sportsgene.ee/genetic-test-of-athletic-abilities.

36. "Optimum Athletic Performance DNA Analysis: Athletic Analysis and Sports Genetic Test," CyGene Direct, 2009, http://cygene.infinityarts.com/browse-10873/Optimum-Athletic-Performance-Dna-Analysis.html.

37. Ibid.

38. S. Salzberg, "Genetic Tests for Kids' Sports Abilities: Hype or Science?," *Forbes*, 2011, www.forbes.com/sites/stevensalzberg/2011/05/21/genetic-tests-for-kids-sports-abilities-hype-or-science/.

39. "Testing," Atlas Sports Genetics, 2011, www.atlasgene.com.

40. Ibid.

41. "Partners," Zybek Sports, 2015, www.zybeksports.com/pages/partners.

42. "Why Do Athletes Take the DNAeX Test?," DNAeX, 2014, http://dnaex.net/exercise-genomics/why-do-athletes-take-a-dnaex-test/.

43. Ibid.

44. "Your CyGene DNA Profile Test Report, Athletic Performance Panel," sample report, CyGene Direct, 2008, http://cygene.infinityarts.com/assets/files/12096.pdf.

45. "About the Test," Genetic Performance, 2014, https://geneticperformance.com/about-the-test-new.

46. Ibid.

47. "AlBioTech Working with FDA to Ensure New Genetic Test Remains Available to Consumers," PR Newswire, 2011, www.prnewswire.com/news-releases/aibiotech-working-with-fda-to-ensure-new-genetic-test-remains-available-to-consumers-122581158.html.

48. "How It Works," SingldOut, 2014, https://singldout.com/#/comingsoon. See also M. Engel, "SingldOut.com Uses Your Genes to Find Your Perfect Romantic Match," *NY Daily News*, 2014, www.nydailynews.com/life-style/health/singldout-genes.

49. Ibid.

50. "About LoveGene," LoveGene, 2014, www.lovegene.co.uk. See also N. Feradov, "A Swab Test to Find True Love? LoveGene Brings Back the 'Chemistry' to Matchmaking," *Renegade Times*, 2014, http://renegadetimes.com/2014/03/31/a-swab-test-to-find-true-love-lovegene-brings-back-the-chemistry-to-matchmaking/.

51. "Relationship Compatibility," Instant Chemistry, 2014, http://instantchemistry.com/relationship-compatibility/.

52. Ibid.

53. "DNA Dating: Finding Your Genetic Match," ABC News, *Good Morning America*, 2009, http://abcnews.go.com/GMA/Weekend/story?id=6921105&page=1.

54. "About Gene Partner," Gene Partner, 2007–15, www.genepartner.com/index.php/aboutgenepartner.

55. "Letters to Manufacturers Concerning Genetic Tests," Food and Drug Administration, 2010, www.fda.gov/MedicalDevices/ProductsandMedicalProcedures/InVitroDiagnostics/ucm219582.htm.

56. "Regulation of Genetic Tests," National Human Genome Research Institute, 2015, www.genome.gov/10002335#al-4.

57. Jennifer Oulette, "23andMe Is Back in the Genetic Testing Business with FDA Approval," *Gizmodo*, 2015, http://gizmodo.com/23andme-is-back-in-the-genetic-testing-business-with-fd-1737917276.

58. A. Pollack, "23andMe Will Resume Giving Users Health Data," *New York Times*, October 21, 2015, www.nytimes.com/2015/10/21/business/23andme-will-resume-giving-users-health-data.html?smprod=nytcore-iphone&smid=nytcore-iphone-share&_r=1.

59. M. Herper, "23andMe Wins a Second Life: New Business Plan Scores $115 Million from Investors," *Forbes*, 2015, www.forbes.com/sites/matthewherper/2015/10/14/23andme-prepares-a-comeback-raising-115-million-at-a-1-1-billion-valuation/.

60. Taha A. Kass-Hout and David Litwack, "Advancing Precision Medicine by Enabling a Collaborative Informatics Community," Food and Drug Administration, *FDA Voice* (blog), 2015, http://blogs.fda.gov/fdavoice/index.php/2015/08/advancing-precision-medicine-by-enabling-a-collaborative-informatics-community/.

61. Jasmine Pennic, "FDA Unveils Open Source Platform for Genomic Sequencing Data," *HIT Consultant*, 2015, http://hitconsultant.net/2015/08/06/fda-open-source -platform-genomic-sequencing-data/.

62. Michael Mezher, "precisionFDA: A Crowd-Sourced, Cloud-Based Platform for Precision," *Regulatory Affairs Professionals Society*, 2015, www.raps.org/Regulatory -Focus/News/2015/08/06/22976/precisionFDA-A-Crowd-Sourced-Cloud-Based -Platform-for-Precision-Medicine/. See also Aaron Krol, "precisionFDA to Test Accuracy of Genomic Analysis Tools," *Bio IT World*, 2015, www.bio-itworld.com/2015/8/27/ precisionfda-test-accuracy-genomic-analysis-tools.html.

63. John Conley, "Conley Q&A on LDTs and the FDA," *Genomics Law Report*, 2015, www.genomicslawreport.com/index.php/2015/01/29/conley-q-a-on-ldts-and-the -fda/#more-13394.

64. "Points to Consider in Assessing When an Investigational Device Exemption (IDE) Might Be Needed," National Human Genome Research Institute, 2015, www.genome .gov/27561291.

65. "Regulation of Genetic Tests," National Human Genome Research Institute, 2015, www.genome.gov/10002335#al-4.

66. "Document Number: GEN1500674 (Letter to Pathway Genomics, Inc.)," Food and Drug Administration, 2015, www.fda.gov/downloads/MedicalDevices/Resourcesfor You/Industry/UCM464092.pdf.

67. "Counsyl Developing Own NIPT; Converting Carrier Screening Test to NGS," *Genome Web*, 2015, www.genomeweb.com/business-news/counsyl-developing-own-nipt -converting-carrier-screening-test-ngs. See also "Counsyl's New Expanded Inherited Cancer Screen Allows Clinicians to Identify Women at Elevated Cancer Risk Who Would Have Otherwise Been Missed," *Business Wire*, 2015, www.businesswire.com/news/ home/20150922005879/en/Counsyl's-Expanded-Inherited-Cancer-Screen-Clinicians -Identify#.VgWo_rTk8zo.

68. "Interleukin Genetics Granted Key European Patent Covering Inherent Health* Weight Management Genetic Test," Interleukin Genetics, 2015, http://ilgenetics.com/ interleukin-genetics-granted-key-european-patent-covering-inherent-health-weight -management-genetic-test/.

69. "Testing," Stanford Sports Genetics, 2010, https://sportsgenetics.stanford.edu/. See also "23andMe Tests NFL Players' DNA for Athletic Genetic Factors," PR Newswire, 2009, www.prnewswire.com/news-releases/23andme-tests-nfl-players-dna-for-athletic -genetic-factors-64084122.html.

70. "Introducing Helix," Helix, www.helix.com/.

71. Jamie Hartford, "Genelex Exec: FDA Too Slow to Regulate LDTs," Medical Device and Diagnostic Industry, 2015, www.mddionline.com/article/genelex-exec-fda-regulation -will-make-personalized-medicine-more-expensive-03-16-15.

72. "Welcome," NHS, UK Genetic Testing Network, 2015, http://ukgtn.nhs.uk.

73. Rebecca Hill, "Human Genetics Commission Publish Final Report," *BioNews*, 2012, www.bionews.org.uk/page_149450.asp.

74. "Genomic Medicine," UK Parliament, House of Lords, 2009, www.publications. parliament.uk/pa/ld200809/ldselect/ldsctech/107/107i.pdf.

75. "Genomics," NHS, Health Education England, 2015, https://hee.nhs.uk/our-work/ hospitals-primary-community-care/genomics.

76. P. Borry, R. E. van Hellemondt, D. Sprumont, C. F. Duarte Jales, E. Rial-Sebbag, T. M. Spranger, L. Curren, J. Kaye, H. Nys, and H. Howard, "Legislation on Direct-to -Consumer Genetic Testing in Seven European Countries," *European Journal of Human Genetics* 20, no. 7 (2012), www.nature.com/ejhg/journal/v20/n7/full/ejhg2011278a. html#close.

77. "Relevant Legislation and Regulatory and Advisory Bodies in the United Kingdom," National Center for Biotechnology Information, Direct to Consumer Genetic Testing: Summary of a Workshop, 2011, www.ncbi.nlm.nih.gov/books/NBK209642/.

78. Michelle Roberts and Paul Rincon, "Controversial DNA Test Comes to UK," *BBC News*, 2014, www.bbc.com/news/science-environment-30285581.

79. Borry et al., "Legislation on Direct-to-Consumer Genetic Testing in Seven European Countries."

80. Ibid.

81. Ian Sample, "Genetic Tests Flawed and Inaccurate, Say Dutch Scientists," *Guardian*, 2011, www.theguardian.com/science/2011/may/30/genetics-tests-flawed-dutch -scientists.

82. Shu-Ching Jean Chen, "China Cracks Down on DNA Testing," *Forbes Asia*, 2014, www.forbes.com/sites/shuchingjeanchen/2014/03/03/china-cracks-down-on-dna -testing-2/.

83. "Berry Genomics Lands China FDA Approval for NIPT Sequencer," *Genome Web*, 2015, www.genomeweb.com/regulatory-news/berry-genomics-lands-china-fda -approval-nipt-sequencer.

84. Robert Cook-Deegan, "Australian Appeals Court Upholds Patents on Isolated BRCA1 DNA," *Genomics Law Report*, 2014, www.genomicslawreport.com/index. php/2014/09/30/australian-appeals-court-upholds-patents-on-isolated-brca1-dna/.

85. "BGI iNex Create NGS Facility in Singapore," *Genome Web*, 2012, www.genome web.com/sequencing/bgi-inex-create-ngs-facility-singapore-offer-molecular-genetic -testing.

86. "German Merck Expanding Presence in Israel with Acquisitions," Reuters, 2015, www.reuters.com/article/merck-israel-idUSL5N0ZF3GF20150629.

87. "Quality Assurance and Proficiency Testing for Molecular Genetic Testing: Summary Results of a Survey of 18 OECD Member Countries," Biotechnology Division, OECD Directorate for Science, Technology, and Industry, Paris, 2015, www.oecd.org/sti/ biotech/34779945.pdf.

88. Ibid.

89. "Direct-to-Consumer (DTC) Genetic Testing: A Global Strategic Business Report," Global Industry Analysts, 2012, www.strategyr.com/Direct_to_Consumer_DTC_ Genetic_Testing_Market_Report.asp.

90. "Global Genetic Testing Market to Reach US$2.2 Billion by 2017, According to New Report by Global Industry Analysts, Inc.," *Biomed Trends*, 2013, www.biomedtrends .com/GetDetails.asp?CatName=Genetic%20Testing.

91. Jason Petron, "For Consumer Genomics Market, 2014 Was a Year of New Features, New Players, and 'Unmatched' Sales," *Genome Web*, 2014, www.genomeweb.com/ microarrays-multiplexing/consumer-genomics-market-2014-was-year-new-features -new-players-and.

92. Pascal Su, "Direct-to-Consumer Genetic Testing: A Comprehensive View," *Yale Journal of Biology and Medicine* 86, no. 3 (2013), www.ncbi.nlm.nih.gov/pmc/articles/ PMC3767220/.

93. Ibid.

94. Justin Petrone, "With 1M Customers Genotyped, 23andMe Continues Consumer Genomics Rebound," *Genome Web*, 2015, www.genomeweb.com/microarrays -multiplexing/1m-customers-genotyped-23andme-continues-consumer-genomics -rebound.

95. "Investors: Annual Reports," Genomic Health, 2014, http://investor.genomichealth .com/annuals.cfm.

96. Timothy Caulfield, "Direct to Consumer Genetic Testing in Canada: Should We Be Concerned?," Healthy Debate, 2014, http://healthydebate.ca/opinions/direct -consumer-genetic-testing.

97. See "Genetic Screening Arrives at the Wellington," GeneHealth UK, 2015, www .genehealthuk.com/news/102-genetic-screening-arrives-at-the-wellington.

98. Petrone, "With 1M Customers Genotyped, 23andMe Continues Consumer Genomics Rebound."

99. Louisa Wilkins, "Genetic Screening," n.d., www.easternbiotech.com/images/ healthbody.pdf. See also "Dubai Biotechnology and Research Park (DuBiotech)," HDR, 2015, www.hdrinc.com/portfolio/dubai-biotechnology-and-research-park-dubiotech.

100. Andrea Downing Peck, "China's Genome Mapping Giant BGI Is Poised to Become an International Leader in Gene Sequencing and May Play Major Role in Interpretation of Genetic Test Results," *Dark Daily*, 2015, www.darkdaily.com/ chinas-genome-mapping-giant-bgi-is-poised-to-become-an-international-leader -in-gene-sequencing-and-may-play-major-role-in-interpretation-of-genetic-test -results#axzz3wj2uGBi3.

101. "Global," Pathway Genomics, 2015, www.pathway.com. See also "Quality and Safety in Genetic Testing," World Health Organization, 2015, www.who.int/genomics/policy/ quality_safety/en/index1.html; Keith Grimaldi et al., "Personal Genetics: Regulatory Framework in Europe from a Service Provider's Perspective," *European Journal of Human Genetics* 19, no. 4 (2011), www.ncbi.nlm.nih.gov/pmc/articles/PMC3060315/; Meredith Knight, "Will FDA Regulations Force US Direct-to-Consumer Genetic Testing Companies Overseas?," *Genetic Literacy Report*, 2014, www.geneticliteracyproject .org/2014/05/14/will-fda-regulations-force-us-direct-to-consumer-genetic-testing-companies -overseas/.

102. "Patrick Chung," Crunch Base, 2015, www.crunchbase.com/person/patrick -chung/advisory-roles.

103. "Esther Dyson," Crunch Base, 2015, www.crunchbase.com/person/esther-dyson #/entity.

104. "Jim Plante, Founder and CEO," Pathway Genomics, 2015, www.pathway.com/ jim-plante-appointed-by-us-secretary-of-commerce-and-us-trade-secretary/.

105. "Dayton Misfeldt," Crunch Base, 2015, www.crunchbase.com/person/dayton -misfeldt#/entity.

106. Rose, *Politics of Life Itself.*

Conclusion

1. Conley and Fletcher, *Genome Factor.*

2. Dorothy Roberts, "Debating the Cause of Health Disparities: Implications for Bioethics and Racial Equality," *Cambridge Quarterly of Healthcare Ethics* 21, no. 3 (2012): 332–41.

3. Duster, *Backdoor to Eugenics.*

4. T. Ishii and A. Motoko, "International Regulatory Landscape and Integration of Corrective Genome Editing into In Vitro Fertilization," *Reproductive Biology and Endocrinology* 12, no. 108 (2014); Stephen Hilgartner, "Mapping Systems and Moral Order: Constituting Property in Genome Laboratories," in Jasonoff, *States of Knowledge*; Paul Knoepfler, "Key Action Items for the Stem Cell Field: Looking Ahead to 2014," *Stem Cells and Development* 22, s1 (2013): 10–12.

5. "International Summit on Human Gene Editing," National Academies of Sciences Engineering Medicine, 2015, http://nationalacademies.org/gene-editing/Gene-Edit -Summit/index.htm. The purpose of the summit was to decide whether or not to permit basic research and/or clinical applications of gene editing at all, especially edits made to the human germ line.

6. "Desktop Genetics Gets £1.3m to Develop Genome Editing Platform," *Cambridge News*, 2015, www.cambridge-news.co.uk/Desktop-Genetics-gets-1-3m-develop -genome-editing/story-27802748-detail/story.html.

Also see "Editas Medicine Raises $120 Million to Advance Genome Editing," *Business Wire*, 2015, www.businesswire.com/news/home/20150810005323/en/Editas-Medicine -Raises-120-Million-Advance-Genome; Cory Renaeur, "Vertex Pharmaceuticals' $2.6 Billion Bet on CRISPR," *Motley Fool*, 2015, www.fool.com/investing/general/2015/11/07/ vertex-pharmaceuticals-26-billion-bet-on-crispr.aspx; Antonio Regalado, "CRISPR Patents Spark Fight to Control Genome Editing," *MIT Technology Review*, 2014, www .technologyreview.com/featuredstory/532796/who-owns-the-biggest-biotech-discovery -of-the-century/.

7. "Genome Editing Market Worth $3,514.08 Million by 2019," Markets and Markets, 2015, www.marketsandmarkets.com/PressReleases/genome-editing-engineering .asp. Also see "BioInformatics LLC Projects Market for CRISPR/Cas9 Products to Grow by 37% in the Next 12 Months," *Business Wire*, 2015, www.businesswire.com/news/ home/20150528006088/en/BioInformatics-LLC-projects-market-CRISPRCas9-products -grow.

INDEX

Page numbers followed by n *refer to notes, with note number.*